Project Management: Novice-To-Expert!

AN EXAMINATION OF PROJECT MANAGERS IN THE ENGINEERING PROCUREMENT AND CONSTRUCTION MANAGEMENT INDUSTRY (EPCM)

Project Management: Novice-To-Expert!

A Qualitative Comparative Case Study

AN EXAMINATION OF PROJECT MANAGERS IN THE ENGINEERING PROCUREMENT AND CONSTRUCTION MANAGEMENT INDUSTRY (EPCM)

Derrick J. Walters, PMP, EdD

Copyright © 2018 by Derrick J. Walters, PMP, EdD.

Library of Congress Control Number:		2018904834
ISBN:	Hardcover	978-1-9845-2349-5
	Softcover	978-1-9845-2348-8
	eBook	978-1-9845-2347-1

All rights reserved. No part of this book may be reproduced or transmitted in any form or by any means, electronic or mechanical, including photocopying, recording, or by any information storage and retrieval system, without permission in writing from the copyright owner.

Any people depicted in stock imagery provided by Getty Images are models, and such images are being used for illustrative purposes only. Certain stock imagery © Getty Images.

Print information available on the last page.

Rev. date: 05/19/2018

To order additional copies of this book, contact:
Xlibris
1-888-795-4274
www.Xlibris.com
Orders@Xlibris.com
769850

Contents

List of Tables vii
List of Figures viii
List of Appendices ix
Abstract xi
Acknowledgements xiii
Dedication xv
Biography xvii

Chapter 1 Introduction 1
    Problem Statement 6
    Purpose 8
    Research Questions 9
    Rationale and Significance 9
    Definitions 10
    Conceptual Framework 11

Chapter 2 Literature Review 16
    Expertise: What is the Nature of it? 17
    Expertise: How is it Developed? 19
    Project Management: A Review of the Profession 22
    Continuing Professional Education: What is it? 29
    CPE Arguments and Contested Issues: What are They? 31
    Informal Learning: What is it? 32
    Informal Learning: Issues and Problems 36

      Comparative Analysis: Project Management is a Relative
      Expertise Gained Through Deliberate Practice 38
      Summary 42

Chapter 3   Methodology 44

      Research Procedure 45
      Data Analysis 49
      Limitations of the Study 51

Chapter 4   Results 53

      Study 1 58
      Study 2 70
      Study 3 86
      Cross-Study Theme Analysis 100
      Assertions and Generalizations 112

Chapter 5   Discussion 115

      Summary of Findings 115
      Relationship between the Research Questions and the
      Findings 117
      Discussion 130
      Contributions to the Literature 134
      Implications for Practice 135
      Suggestions for Future Research 137

References 139

List of Tables

Table 1	Problems Faced by Project Management Companies	27
Table 2	Characteristics of Formal Learning, Informal Learning and Incidental Learning	35
Table 3	Interview Process Timeline	46
Table 4	Participant Demographics	48
Table 5	Code to Theme	55

List of Figures

Figure 1 Figure 1. Conceptual Framework. 14
Figure 2 Figure 2. Portrait of three studies. 50
Figure 3 Figure 3. Conceptual framework updated (Walters, 2015). 54

List of Appendices

A	Interview Questions	149
B	Invitation To Participate	155
C	Informed Consent Form	159
D	Demographics Form	165
E	Code-To-Theme Tables	169
F	Member Check - Followup Email	175

Abstract

THE ENGINEERING PROCUREMENT and construction management (EPCM) industry in the Chicagoland area is unique in that its customer base is primarily made up of oil and manufacturing companies. This uniqueness extends to the project managers in the EPCM industry, in that, a large percentage of the project managers who enter the industry with the title project manager, or get promoted to project manager, don't have a clear understanding of the processes or steps required to achieve project management expertise. As a result, little is known about the path of project managers and how they achieve the outcomes that characterize the project management achievement levels – novice, intermediate, or expert – that designate their professional achievement.

The purpose of this study was to explore the project management domain — or area of knowledge — in the EPCM industry and the expertise that underlies and delineates a project manager's competencies (achievements) — novice, intermediate, and expert — for the purpose of understanding how these competencies (achievements) are accomplished in the project management domain.

I used three studies in a qualitative method to investigate project managers' growth from novice to expert in the project management domain. Three project managers with varying project management experiences, were asked to share their stories, experiences, and opinions in the context of the EPCM industry.

The results of this study were revealing; each of the three project managers used different and similar methods to negotiate the novice to expert path, while achieving varying measures of success in their project management careers.

Acknowledgements

I WOULD LIKE TO express my sincere appreciation to Dr. Wei-Chen Hung, for his staunch commitment to lead, coach, counsel, and challenge me throughout this entire research process. He went above and beyond the call of duty to make sure I stayed the course and remained focused in the face of adversity. I would also like to thank Dr. Jorge Jeria and Dr. Terry Borg for their insight, support, and expertise. They were especially instrumental in helping me fine tune my research and transfer my thoughts into manuscript.

Lastly, I would like to thank Dr. Gene Roth who begin this research journey with me but was unable to complete it because of personal commitments and other initiatives that were more pressing and fulfilling. His firm but fair approach was greatly appreciated in the early stages of my research.

DEDICATION

THIS RESEARCH STUDY and resulting literature is dedicated to my wife, Pamela, whose patience, support, love, and unwavering tenacity helped me stay focused while completing this project. You deserve additional credit for being my biggest cheerleader as well as my trainer and corner-person. It would take an eternity to explain how much your support has meant to me during this arduous journey. You are to be saluted and commended for propping me up and telling me to keep on keeping on.

To my daughters: Felicia, Delicia, Gabrielle, Joselyn and Jasmine anything is possible if you put your mind to it, follow my lead.

To my late Pastor, Apostle Richard D. Henton, who poured God's word into me, and gave me the foundation required to squeeze every ounce of potential from my mortal frame; I thank you sir. I will always remember two of your favorite sayings: (1) Faith is a leap into the darkness, with the expectation of a safe landing. (2) Faith is leaping into a wall looking for the hole to be there when you get there. I will see you again.

And finally, to my dad, Apostle Lynnell Walters Sr.; you get credit for sending me on a journey that required me to never give up on my dream. I will forever remember the day you uttered the words: "Derrick (son) the time it takes to complete your education will pass, if you live, it will pass with you thinking about it, or it will pass with you doing it, make your choice!"

Biography

DR. DERRICK WALTERS has a diverse background and resume that includes 30 years of maintenance, engineering, business, management, project management, university teaching, corporate training, and consulting experience. He has held several middle management positions in engineering, maintenance, and business with progressive responsibilities that have allowed him to enhance and sharpen his project and business acumen. He has been the lead project manager for projects that range from $100K to $200M, and his project teams typically consist of 10-15 cross-functional professionals with advanced degrees in engineering: Civil, Structural, mechanical, and electrical. Also included were maintenance, management, computer networking, database administrators, developers, process engineers, stress analyst, architects, designers, and a team of administrative personnel.

Professor Walters completed his undergraduate degree in engineering technology at Purdue University, his MBA in project management at Keller Graduate School of Management, and his doctoral studies (EdD) at Northern Illinois University (NIU). In addition, he completed the requirement for PMI's PMP certification in 2005, and has been an active PMP for 13 years.

Dr. Walters currently lives in the suburbs of Chicago, IL and owns a consulting company, Walters Consulting, LLC. He has been teaching PMP Prep courses for the past 13 years, while teaching at colleges and universities (online and on-ground) for the past 17 years. In addition, he advocates the use of computer simulations to illuminate and reinforce topics like project management. His interests include NBA basketball, NFL football, ML Baseball, Boxing, the UFC, Gospel music, the Internet, video games, and reading.

Chapter 1

INTRODUCTION

I WAS EMPLOYED FOR approximately 10 years as a project manager in the engineering procurement and construction management (EPCM) industry. Four of those years, 2005–9, were spent working at a company that I will refer to as Company A. Much of the information provided in this introduction is based on what I learned while working as a project manager in the EPCM industry. The EPCM industry is unique in that it is comprised of professional engineering companies that supplement the workforce of local companies for the purpose of planning and executing project work that companies throughout the Chicagoland and northwestern Indiana areas are not capable of doing, don't have the resources to do, or don't have the time to do; most EPCM companies in the area have expanded their business and are doing business in other states and some countries. The primary customers of the EPCM industry are based in the oil and manufacturing industry. In addition, because the EPCM industry performs contract work and employs project managers, engineering professionals, architects, designers, and drafters in a consultancy-type format, they basically sell work hours. Therefore, if they are not doing project work (selling work hours) for other companies, their net profit may be affected. Consequently, to maintain a positive financial position, they try to bill or assign their professionals to customer project work, or else risk incurring the cost of their professionals on a daily basis as overhead cost. As a result of this profit-loss structure, the majority of companies within the EPCM industry are ill-prepared to withstand prolonged, increased overhead cost. Therefore, companies in the EPCM industry, as well as the project managers that work for them, were (and are) encouraged to pursue project work for their respective companies.

As a project management consultant in 2008, I witnessed a shortage of project managers (PMs) in the local EPCM industry. This shortage became more apparent when one of the oil companies in Northwest Indiana, British Petroleum (BP), announced a ten-year, $5 billion upgrade initiative that would require the EPCM industry to enhance their workforce by hiring more PMs. This announcement sent the companies in the EPCM industry scrambling to acquire talent that would position them to take advantage of the project management needs created by BP's upgrade and expansion initiatives. Unfortunately, many EPCM companies in the area were pulling from the same pool of resources, essentially magnifying the problem, creating more of a shortage, and causing more consternation. To compound the problem, a few project managers moved from company to company, working over a two-year period for several EPCM companies in the area. This created instability and knowledge-transfer issues that weakened companies and affected project continuity.

More importantly, BP's upgrade announcement and its ensuing need for project managers shined the light on an already festering problem, which was a shortage of qualified project managers with the skills needed to deliver projects on time, under budget, and within quality standards. An executive vice president at Company A told me that they were in need of talent, and they likened themselves to a bus driving down the street, picking up everyone who was qualified. Moreover, in an effort to acquire talent, Company A introduced an employee referral program that rewarded its employees, including me, for recommending qualified candidates who were hired and worked for the company a minimum of ninety days.

Fast forward to June of 2011, and this same company, Company A, appeared to have the same problem. In a conversation I had with a general manager at Company A, I was informed that Company A was expanding, so the need for talent still existed. With the shortfall in talent and the desperate needs of the oil industry, one would think consulting companies in the EPCM industry would be working to perfect their project management trade and shore up the weaknesses that hinder project management performance and timely project delivery. Unfortunately, that was not the case; newcomers to the field of project management were amazed at the number of projects that were delivered late, over budget, and outside of quality standards.

However, those in the EPCM industry were familiar with the problems that plague project management, such as projects that are late, over budget, and outside of quality standards. In addition, according to the Standish Group (2009), many companies were losing money because their project managers were ill-prepared to lead and manage company initiatives (projects) that require meticulous preparation, planning, and execution.

In March of 2013 - 14, I went back into the EPCM industry to work as a project manager for a short period of time, and I discovered the problems that plagued the industry in 2008 and 2011 were still very apparent: There was still a shortage of qualified project managers, and so much so that one of the project managers that I worked alongside came to the Chicagoland area from the East Coast to fill a project management need (position).

He planned to work in the area for a short period of time, complete his assigned project work, and then move back to the East Coast to be with his family.

Furthermore, in that March 2013-14 timeframe, I discovered that some of the project managers still struggled to understand some of the basic project management jargon and concepts that characterize the project management profession. More importantly, it has been my experience that not understanding these items can lead to, and usually does lead to, negative project results.

In August of 2014, I was contacted by a recruiter and offered a long-term project assignment in St. Louis, Missouri—"long term" meant three years. The recruiter stated that there was a shortage of project managers who were willing to relocate to St. Louis, and unfortunately for the recruiter and the oil company that submitted the requisition, I was not willing to relocate to St. Louis for three years either.

Another problem that still exists, according to an executive at another company, Company B, a competitor of Company A, is the absence of a vehicle in the form of experiential learning, a training program, or some other formal mechanisms that will help a project coordinator, project engineer, or senior project engineer transition to the position of project manager. Without these items in place, companies could continue delivering projects that are subpar, while perpetuating an environment of mediocrity. This type of environment will prolong the project management learning curve and increase the distance between where we are now and the plateau that represents regular project management success.

It has been suggested that future experts gradually acquire patterns and knowledge about how to react in situations by storing memories of their past actions in similar situations (Simon and Chase 1973). Therefore, I conclude that Simon and Chase meant performance could improve through renewed and continued experiences. I believe project management performance can improve through continuing professional education (CPE), formal and informal training, the acquisition of project management professional (PMP) certification, and renewed and continued experiences.

Galton acknowledged and pointed out the need for and the importance of training and practice as a mechanism to reach high levels of performance in any domain (1979, p684), and this includes the domain of project management.

The problems described in the EPCM industry are not reserved or indigenous to the EPCM industry alone; similar project management–related problems exist in the information technology (IT) industry. For example, approximately 70 percent of all IT projects are delivered late and over budget, while 52 percent of all IT projects come in at 189 percent of budget, and a big percentage of IT projects are canceled because of poor preparation or poor planning (Standish Group 2004).

Sadly, IT project studies conducted by the Standish Group (2009) show that approximately 70–90 percent of all project initiatives fail. In addition, according to Padgett (2009) these are the types of results that progressive companies would like to improve on, if not eradicate. Consequently, the data provided by the Standish Group and Padgett provides some proof that the project management profession has issues that need to be addressed.

According to Ericsson, there are several factors that influence our level of professional achievement and ultimately affect whether or not a person reaches the level of expert in his or her profession. One factor is extensive experience of activities in a domain (2006, p683).

Problem Statement

A preliminary and somewhat superficial look at project management in the EPCM industry indicates a mixed bag of problems. According to senior executives at a Company called A (personal communication, December 3, 2014), the overarching and main problem is that there is not a clear definition of the processes or steps required to achieve expertise or the status of expert project manager in the EPCM industry. I believe this problem is compounded by the lack of a formal or informal training program that will enable or assist novice project managers to smoothly transition, grow, or evolve into expert project managers. Unfortunately, the EPCM industry has not adopted PMI's certification programs in project management. In addition, an executive vice president at Company A indicated in an interview with me that he would not require his project managers to pursue certification unless his customers required it.

Moreover, these same executives say there appears to be three conditions contributing to the main problem. The first of these three conditions points to a lack of appreciation for CPE in the EPCM industry. For example, in the 2014 interviews I conducted with four executive vice presidents of three local EPCM companies, three of the four vice presidents did not allocate funds for training or professional development for their project managers. The one vice president who did allocate funds, from Company B, admitted the money allocated was a small amount and not nearly enough to have a significant effect.

The second condition is a belief that CPE does not translate into measurable or increased job performance. Therefore, it is hard for anyone in the EPCM industry to embrace or justify it. Unfortunately, the belief that CPE does not translate into measurable or increased job performance is not a new revelation. Nowlen argued few providers of CPE can demonstrate that specific educational programs affect practitioners' performance or enhance competency, therefore care must be taken that continuing professional educators do not purport to accomplish something they cannot deliver (1988, p232).

The third condition has to do with work hours. The EPCM industry sells work hours, and every work hour that CPE takes away from daily work hours, or billable hours, is transferred to non-billable or overhead hours. According to an executive from company B, this presents a double whammy (personal communication, December 3, 2014).

More importantly, if no one is willing to address the problems mentioned above, project managers may experience difficulty achieving the level of project management expertise that is required for successful project delivery. As mentioned earlier, Ericsson (2006) suggested that there are several factors that influence the level of professional achievement or expert status. Those factors include experience, training, formal education, supervision by more experienced professionals, and regular, prolonged execution of activities that characterize the professional domain that they are working in (p. 683). Subsequently, after months, possibly years, of carrying out the activities they've learned, they could reach an acceptable level of proficiency that could lead to the coveted status of expert.

However, relative to this study, little is known about the path of project managers and how they achieve the outcomes that delineate and characterize the titles – novice, intermediate, or expert – that were loosely alluded to by Ericsson. I hope to shed some light on this topic during this study.

Purpose

The purpose of this study was to explore the project management domain — or area of knowledge — in the EPCM industry and the expertise that underlies and delineates a project manager's competencies at the novice, intermediate, and expert levels, for the purpose of understanding how these outcomes are achieved in the project management domain. I initiated and completed the studies through a qualitative approach to investigate project managers' growth from novice to expert in the project management domain. In addition, I researched the problem and the conditions surrounding the problem mentioned earlier in the study to determine the activities that would enable or assist novice project managers in their quest to smoothly transition, grow, and/or evolve into expert project managers in the EPCM industry.

Previous research (Benner, 1982, 1984; Benner & Tanner, 1987; Dreyfus & Dreyfus, 1985) identified five stages of professional development: novice, advanced beginner, competent, proficient, and expert; while Daley's (1999) research focused on two stages, a comparison between novice and expert. Daley's (1999) study analyzed the different learning processes undertaken by novices and experts. Specifically, Daley conducted semi-structured interviews with nurses for the purpose of analyzing and comparing how nurses' learning developed or was developed in their clinical practice. My study is based on Daley's work, in that I investigated the same novice-to-expert continuum of professional practice development; however, I explored project management, whereas Daley explored nurses. Therefore, I used Daley's research focus as a guide.

The next section describes the research questions for this study. I will discuss the research questions in Chapter 5.

Research Questions

1. How do project managers (PMs) in the EPCM industry conceptualize and define the role of an expert project manager?
2. How do project managers describe their evolution from novice to expert?
3. Are there any other factors that contribute to project managers becoming experts? If so, how are these factors related to the three types of expertise: absolute, relative, and the theory of deliberate practice?

Rationale and Significance

The rationale behind my selection of this topic stems from what I believe is a paucity of knowledgeable and competent project managers who rely on a knowledge base of recognized tools, techniques, and methodologies to produce projects that are congruent with the "triple constraints": scope, time, and cost.

According to Gray and Larson (2006), "quality and the ultimate success of a project are traditionally defined as meeting or exceeding the expectations of the customer and/or upper management in terms of performance (scope), schedule (time), and budget (cost) of the project" (p.103). On a regular basis, and as stated by the Standish Group, many projects, typically led by project managers, struggle to produce these required outcomes.

In the field of project management, some expert project managers develop and grow into their roles without the benefit of CPE and/or certification programs. Little is known about this select group of project managers and how they developed their knowledge and gain their expertise. This study is extremely significant because the results of this study could provide the EPCM industry with answers about this select group of project managers, while providing the EPCM industry with the blueprint they need to create a formal initiative and / or agenda that explains how to achieve the level or status of expert project manager.

Definitions

Continuing professional education (CPE) – Continuing professional education, by definition, refers to the education of professional practitioners, regardless of their practice setting, that follows their preparatory curriculum and extends their learning throughout their careers (Queeney, 1996).

Continuing professional development (CPD) – Continuing professional development, outside of the United States, is a broad category that encompasses formal and informal learning. But in the United States, continuing professional education is the more commonly used term (Jeris, 2010 p. 277).

Domain – An area (body) of knowledge or activity.

Expert – A distinguished or brilliant journeyman, highly regarded by peers, whose judgments are uncommonly accurate and reliable, whose performance shows consummate skill and economy of effort, and who can deal effectively with certain types of rare or "tough" cases (Chi, 2006, p. 22).

Informal Learning – Informal learning is learning that takes place in the normal course of daily events without a high degree of design or structure and occurs from and through experience (Marsick & Watkins, 1990, p. 15).

Novice – A novice is literally someone who is new, a probationary member. Someone with minimal exposure to the domain (Chi, 2006, p. 22).

Project – A temporary endeavor undertaken to produce a unique product service or result (Project Management Institute [PMI], 2016).

Project Management – The application of knowledge, skills, tools, and techniques to project requirements to bring projects in on time, under budget, and within quality standards (PMI, 2004).

Project Management Institute (PMI) – PMI is the world's leading not-for-profit membership association for the project management profession.

Project Management Professional (PMP) – PMI's Project Management Professional credential is an industry-recognized certification for project managers. Globally recognized and demanded, the PMP demonstrates that you have the experience, education, and competency to successfully lead and direct projects.

Professional Development Unit (PDU) – PDU is a unit of measurement that provide a PMP with one professional-development unit for each PMI-approved hour of training that a PMP attends.

Registered Education Provider (REP) – PMI REPs are organizations that have been approved to offer training in project management and issue professional development units (PDUs) to meet the continuing education requirements needed by PMI credential holders.

Conceptual Framework

According to Merriam (1998), qualitative research attempts to find out how people make meaning or interpret a phenomenon. The conceptual framework for this study includes an examination of project managers in the context of the EPCM industry. I interviewed three project managers (PMs), whose experience levels range from 5 to 30 years in the field of project management. The PMs were selected through convenient sampling from a pool of project managers that I encountered over the past 25 years.

The project managers selected had the following characteristics: The first PM had formal project management education and is a Project Management Institute (PMI)-certified Project Management Professional (PMP); the second PM had formal project management education, but is not PMP certified; and the third PM had no formal education and is not PMP certified. In addition, at least one of the three persons interviewed is a female project manager.

Further, this study relies on four strands of literature: 1) expertise, 2) project management, 3) informal learning, and 4) continuing professional education. Each of these strands was examined individually in an effort to help me understand how project managers make the transition from novice to expert. I will begin with a discussion on expertise, continue with a discussion on project management, followed by continuing professional education (CPE), and then close with the informal learning activities that are typically used by project managers within their professions to move from novice to expert.

There have been studies and research on expertise, CPE, informal learning, and how professionals in other professions make the transition from novice to expert, but little is known about the steps, path, or journey of project managers from novice to expert (Daley, 1999). Figure 1 provides a pictorial look at this study's conceptual framework.

The purpose of this study was to explore the project management domain — area of knowledge — in the EPCM industry and the expertise that underlies and delineates a project manager's competencies at the novice, intermediate, and expert levels, for the purpose of understanding how these outcomes are achieved in the project management domain. I used three studies in a qualitative approach to help with the research. After isolating expertise as the basis of my study, I subsequently identified the different types of expertise using the bin (construct) approach and settled on three types of expertise (bins) that could be investigated as a foundation for the project management domain: 1) absolute expertise, 2) relative expertise, and 3) the theory of deliberate practice.

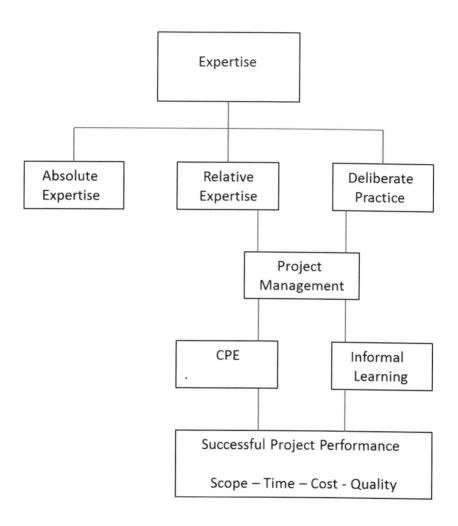

(Walters, 2005)

Figure 1. Conceptual Framework.

Absolute expertise (Simonton, 1977) and relative expertise (Chi, 2006) are two distinct forms of expertise, while the theory of deliberate practice, according to Ericsson (2006), is a method of expertise that is used to improve your domain performance, in this case the domain being project management.

In the absolute domain of expertise, there is a tacit assumption that greatness or creativity arises from chance and unique innate talent (Simonton, 1977). However, because of the nature of project management and how it is defined, project management could not be achieved by chance or the innate means defined by Simonton. According to the PMI (2005), project management is the application of knowledge, skills, tools, and techniques to project requirement to bring projects in on time, under budget, and within quality standards. In order to apply knowledge, skills, tools, and techniques, a project manager must first acquire them. They cannot be acquired by a project manager in the absolute domain because this domain represents chance and innate methods. The focus for this group is on performance. Absolute experts improve their performance with experience (Simonton, 1977). Whereas the focus for those in the relative expertise group focus on understanding how to enable a less skilled or experienced person to become a more skilled person (Chi, 2006).

The other important aspect of this study is project management. This study looks at expertise in the project management domain and then investigates the bins or constructs of CPE and informal learning as vehicles that can be used to help project managers in the project management domain achieve successful outcomes. Successful outcomes or successful project performances in project management are defined as projects that are completed within scope, on time, under budget, and within quality standards.

Chapter 2

LITERATURE REVIEW

This chapter explores the four strands of literature associated with this study's conceptual framework. The four strands of literature are listed below in the order that they will be examined in this chapter:

- Expertise
- Project management
- Continuing professional education (CPE)
- Informal learning

In this research and analysis, I review, synthesize, and critique these strands of literature as a precursor to additional work that led to central research questions, conceptual framework, and proposition of a method for my dissertation.

Expertise: What is the Nature of it?

Theories of expertise and studies of its development have moved through two generations (Holyoak, 1991). The first generation, according to Newell and Simon (1972), was short-lived and centered on expertise as a serial problem-solving that could be applied across a wide range of knowledge and professional domains. The second generation studies examined the precise professional development activities of physicists (Chi, Feltovich, & Glaser, 1980), pilots (Dreyfus & Dreyfus, 1980, 1985), and nurses (Benner, 1982, 1983, 1984; Benner & Tanner, 1987; Tanner, Padrick, Westfall, &Putzier, 1987).

These studies, according to Daley (1999), demonstrated that professionals grow in their chosen career as they gain experience within the context of their work setting.

For example, the studies that were completed in the physics profession showed that novice and experts physicists analyzed problems differently. Findings indicated that "experts initiate abstract physics principles to approach and solve a problem, whereas novices base their approach to a problem on the problem's literal features" (Chi et al., 1980, p.121).

According to Merriam Webster's Dictionary (1999) the definition of expertise is the skill of an expert, where an expert is a person that has special skills or knowledge representing mastery of a particular subject. In addition, the word expert is defined a second time in Merriam Webster's dictionary as, "having, involving, or displaying special skills or knowledge derived from training or experience" (p. 409). According to Webster's New World dictionary (1968), an expert is "one who is very skillful and well-informed in some special field" or "someone widely recognized as a reliable source of knowledge, technique, or skill whose judgment is accorded authority and status by the public or his or her peers" (p.168).

Expertise refers to the characteristics, skills, and knowledge that distinguish experts from novice and less experienced people (Ericsson, 2006). Colvin (2008) suggested that expertise is the ability to do something better than most people can do it.

According to Larson (1984):

> The question to ask about expert power today is not whether the command of esoteric knowledge is a new source of political legitimacy, for it has always been that. Rather, in view of the "postindustrial" trends that are unfolding in all the advanced economies, the question to ask is whether the possession of scientific and technical knowledge can now directly confer political power upon its possessors. (p. 28)

A critical review of Larson's comments could yield the following: expertise is the condition, act, or situation in which a person possesses scientific and technical knowledge.

Contreras (2007) suggested that experts and/or expertise is not a function of education, but rather a function of how well you do research, as in data collection, and how well published you are (p. 5). Specifically, publishing in a discipline's major journals, along with the quality of your data collection, qualifies you as an expert. He seems to indicate a disdain for education alone as a prerequisite for the coveted title of expert, but recommends and advocates for experience and recognition by the field in question as the definitive test or measure of an expert or expertise.

A substantial body of work and research has been done on expertise and how it is developed. This research has produced a handbook on the topic, and the major contributors over the past several decades have been

Anderson (1981); (Bloom) 1985; Chase (1973); Chi, Glaser, and Farr (1988); Ericsson (1996); Ericsson and Smith (1991); Feltovich, Ford, and Hoffman (1997); Hoffman (1992); Starkes and Allard (1993); and Starkes and Ericsson (2003). As a result, there are as many diverse opinions on the development of expertise as there are the nature of expertise.

Expertise: How is it Developed?

Absolute Expertise

In the absolute domain of expertise, there is a tacit assumption that greatness or creativity arises from chance and unique innate talent (Simonton, 1977). Therefore, absolute expertise in this domain is not necessarily developed. Specifically, expertise in project management cannot be developed under the absolute expertise banner, based on Simonton's comment. Instead, the focus for this group is on performance. Absolute experts improve their performance with experience. An example of this would be a 10-year-old chess player achieving grand master status. In the chess world, achieving grand master status at such a young age is an aberration; however, you would still expect the 10-year-old child to get better with more experience.

Twenty-nine years after Simonton's statement about expertise in the absolute domain, Chi (2006) stated that, in the absolute domain, there are several methods that can be used to identify an expert in a particular field. The first method is retrospective, the second is concurrent measurement, and the third is the use of an independent index. Chi (2006) goes on to say:

The retrospective method looks at the outcome or product of an expert to determine how well it was received. For example, to identify a great clothes designer, a person would examine how well his/her clothes sell over a period of years. The second method, the concurrent measurement method, measures performance through some type of rating system. An example of this would be the rating system used in chess tournaments. The third method might be the use of an independent index, an independent tool or assessment in which a person carries out a single task and that task is used as a tool to measure performance. The results is then used to determine if an individual is or will be categorized separately from the masses, as it relates to absolute expertise. (p.21)

Relative Expertise

In the relative approach to expertise, a goal is to understand how we can enable a less skilled or experienced person to become a more skilled person, since the assumption is that expertise can be attained by a majority of students (Chi, 2006). Moreover, relative expertise assumes a separation and a comparison between those who are experts and those who are not; with those who are not given the title of novice. Additionally, the relative approach to expertise assumes that expertise is a level of proficiency that the novice can achieve. However, studies show that a large, organized body of domain knowledge is a prerequisite to expertise and, by definition, experts possess a greater quantity of domain-relevant knowledge than novice (Chi, 2006).

Consequently, it could be stated that a novice would need a sustained level of effort, targeted experience, and expert support to reach the level of expert. There are several methods that can be used to measure proficiency in this relative domain of expertise: 1) academic qualifications, 2) seniority or number of years performing the task, or performance test (Chi, 2006).

One of the advantages of the relative expertise domain is that a less precise definition of the term expertise is needed because we are comparing experts to novice on a continuum that has fewer boundaries. As a result, there is more flexibility and more room for opportunity to set policy, structure training, provide guidelines, and affect outcomes of novices and those who are between the proficiency levels of novice and expert. According to Ericsson (2005), our evolving and keener understanding of what distinguishes experts from novice should lead to better and more effective training (p. 235). The next section on deliberate practice illuminates the specific steps that can be leveraged to further develop one's expertise.

Theory of Deliberate Practice

The goal of deliberate practice is to improve specific aspects of performance in such a way that attained changes or level of proficiency can be measured and integrated into a person's performance (Ericsson, 2006, p. 698). The core assumption of deliberate practice is that expert performance requires the meticulous search for training and tasks that a person can master sequentially.

In addition, deliberate practice presents a person with tasks that are initially outside his/her current realm of performance. As a result, this domain of expertise can be characterized as a developmental domain that has promise as it relates to project management and other performance-based professional disciplines. The inherent nature in which deliberate practice unfolds has a similar approach to that of an apprenticeship method of teaching and learning.

Research on deliberate practice shows that consistent, continued attempts for mastery require those who aspire to expertise, to stretch performance beyond current boundaries or capabilities to compensate for some weakness while preserving those aspects of performance that have already been mastered (Ericsson, 2006, p. 698).

Ericsson (2006) goes on to say:

> Until most individuals recognize that sustained training and effort is a prerequisite for reaching expert levels of performance, they will continue to misattribute lesser achievement to the lack of natural gifts (absolute expertise), and will thus fail to reach their own potential. (p. 699)

Ericsson appears to admonish the majority of people to pursue a targeted, precise, and focused training agenda that will help them reach their potential. Based on previous information, gathered from investigation of absolute and relative expertise, the theory of deliberate practice appears to be a better approach to developing expertise. In addition, those that find themselves in the absolute domain can benefit from performance stretching concepts that are inherent in the theory of deliberate practice.

Project Management: A Review of the Profession

In project management, project managers complete company initiatives that are typically given the name project. In order to fully understand project management, I must first look at the definition of a project. The word project is frequently used in textbooks and standards, but rarely is it accompanied by a precise definition of the concept (Munk-Madsen, 2005). One reason for this is people struggle with the difference between projects and operations. A project has certain characteristics that set it apart from operations (or repetitious activities).

Projects

According to the project management body of knowledge (PMBOK), a project is a temporary endeavor undertaken to produce a unique product, service, or result (PMI, 2004, p. 5). Typically, there are several characteristics that delineate projects from operations. Projects: 1) are temporary, meaning, they have a definite start and end date; 2) are cross-functional in their use of resources; 3) have a predefined budget; 4) are progressively elaborated; and 5) are non-repetitious.

Nicholas and Steyn (2008) stated that, "a project is unique and unfamiliar in some sense and requires multi-functional or multi-organization involvement" (p. 6).

Kerzner (2009) stated,

> a project can be considered any series of activities and task that: have a specific objective to be completed within certain specifications, have a defined start and end date, have funding limits, consume human and nonhuman resources (i.e., money, people, equipment), and are multifunctional (i.e., cut across several functional lines). (p.2).

DeCarlo (2004) advanced the definition of a project to a level that has not been mentioned in the project management body of knowledge. He suggested that a "project is a localized energy field comprising a set of thoughts, emotions, and interactions embodied in physical form" (p.31).

Ekstedt, Lundin, Soderholm and Wirdenius (1999) stated, "a project is a major and significant endeavor or task to be fulfilled within a specific amount of time and with a set number of resources." They went on to state:

> In the case of projects, it is correct to say that the notion of action is almost part of the very definition of a project, consisting of project task, time delimitation for a project, allocation of resources (such as forming a team), and transition (in terms of project progression). (p. 458)

Projects are managed by project managers, and a project manager's main concern should be delivery of projects that are consistent with the triple constraints of scope, time, and cost. A clear understanding of what a project is, coupled with experience and an adequate level of knowledge about the tools and techniques that can be used to deliver a project, is a good basis for timely delivery of projects that are under budget and within quality standards.

My working definition of a project comes from the PMBOK: a project is a temporary endeavor undertaken to produce a unique product, service, or result. After understanding what a project is, we can now delve into and get a definitive understanding of project management (discussed in the next section).

Project Management

The origin of project management has been associated with the space program of the 1960s, but as you can probably imagine, its origin goes back much earlier. The first project managers built the Egyptian pyramids and the Great Wall of China (Wysocki, Lewis, & DeCarlo, 2001). Since the 1960s, project management has made a slow but steady climb to prominence, and today it is an essential component and a necessary tool for companies who want to strategically position their companies for success and remain a viable force in their respective industries.

Further, project management is the endeavor by which a project or group of activities are brought to a successful conclusion. Typically, a project should have three components: (1) clear objectives embedded in a project scope that is linked to an organization and are based on quality, cost, and time; (2) a management process that includes planning, organizing, implementing, and controlling; and (3) a focus that has all levels in the organization, both strategic and tactical, addressed (Turner, 1993).

According to Kerzner (2009), "project management has evolved from a management philosophy restricted to a few functional areas, and regarded as something nice to have, to an enterprise project management system affecting every functional unit of the company" (p. 2).

Simply stated, project management has evolved into a business process rather than a project management process. It is no longer considered an option, but rather a mandatory tool for the survival of most companies.

The Association for Project Management (2000) defined project management as "planning, organizing, monitoring, and controlling of all components of a project, while motivating those involved to achieve project objectives in a safe manner and within the negotiated scope, time, and cost" (p. 5).

According to Gilley and Cunich (1998), project management is a method and set of techniques based on the accepted management principles of planning, organizing, directing, and controlling. Each of these principles is used in combination to reach a desired results on time, within budget, and within quality specifications. I disagree with the notion that project management and management are linked or similar; they are two distinctly different disciplines. A project manager is a leader who embraces changes, manages change, and is constantly dealing with adversity and conflict. These items are inherent in every project.

Management, on the other hand, is different. According to Robbins and Coulter (2012), a manager carries out four functions: plan, organize, lead, and control (p. 9). A close look at the two disciplines (project management and management) indicates requisite and similar levels of skill and knowledge, as well as dissimilar skill sets. A major difference between the two is managers are not managing unique, one-time endeavors that include human resources who are assembled for the sole purpose of completing a project. There are certain dynamics associated with this alone that make management different from project management.

According to the PMI (2005), project management can be defined as the application of knowledge, skills, tools and techniques to project activities to meet project requirements (p. 5). For the purpose of this study, project management is the application of knowledge, skills, tools, and techniques to project requirement to bring project in on time (time), under budget (cost), with minimum deviation from scope, and within quality standards. The focus should be on the triple constraints of scope, time, and cost and, if it is not, the project manager and the project team are focusing on the wrong initiatives.

In addition, according to Pandya (2014), project managers are liable for the day-to-day supervision of the project, specifically the triple constraints of scope, time, and cost. However, they are also responsible for managing change, resource readiness, conflict resolutions, and emotional flare-ups with internal and external stakeholders, and relationship building that could aid in the creation of a high performance team. Dubois, Koch, Hanlon, Nyatuga, and Kerr (2015) stated that project management is an evolving practice and the project manager, leading the projects, has the important role of overseeing the project and project team and, ultimately, ensuring the project ends in success.

Project Management Issues

According to Wilkinson (2001) the project management issues in the EPCM industry can be summarized by the items listed in Table 1. There were 23 companies in the EPCM industry surveyed and Table 1 shows the results of that survey. The first column in Table 1 lists problems that exist in the EPCM industry; the second column, comments, is an explanation of the specific problems encountered; the third column is the number of companies that experienced the problem listed; and the last column shows the percentage of the 23 companies surveyed that experienced the problem listed in column 1.

According to Shen and Yu (2013), many of the issues in project management are related to the processes and limitations of the current practices in project management, the lack of a practical framework, misinterpretation of requirements, difficulties in identifying requirements, conflicts between expectation and constraints, the complex hierarchy of a client's organization, and communication problems in eliciting client requirements (p.223).

Alexander and Stevens (2002) suggested the lack of effective communication contributes to the issues and problems that consistently plague project management. Specifically, project managers must become better communicators and understand the importance of clear, effective communication between groups of project stakeholders who may never meet and who may have dissimilar beliefs and opinions.

Table 1
Problems Faced by Project Management Companies

Problem	Comments	#/23	%
PM lacks respect for others	PM viewed as: Lacking engineering knowledge Expensive parasite at times A paper pusher Disregards input Fails to identify key deliverables	15	65
Client-related problems	Client politics Obtaining client information Poor client information Failure by client to align scope and budget Managing client expectations	11	48
Communication problems	Intermittent communication breakdown Timing of work onsite Equipment not bought early enough Lack of communication	4	18
Being paid on time	Getting paid on time Money	2	8
Other problems	Managing risk Local authorities Subcontractors Quality Set fee tends to minimize workload	5	22

Other project management issues include a lack of documentation on changes, a lack of feedback on requirements, and a lack of well-documented updates make it difficult to track changes in client requirements. Changes to requirements should be documented and tracked, updated and recorded for future reference (Oberg, Probasco, & Ericsson, 2003).

Clarke (2010) suggested that project managers' inability to control their emotions is an issue. Project managers are consistently subject to emotion-generating situations during project management, and their emotional awareness plays a part in determining how they respond to the emotional information generated. Ultimately, these emotions were found to play a significant role in the decisions and behaviors of a project manager, which could impact project outcomes.

Continuing Professional Education: What is it?

Continuing professional education (CPE) should be an integral component of the training and professional development agenda of professionals in general and those who work in the field of project management in particular. Professions and professionals are synonymous with or closely related to professionalization. Professionalization includes three key elements: knowledge base, graduate education, and professional associations (Imel, Brockett, & James, 2000). According to Daley (1999) the connection between learning and the development of practice, or professionalization is an issue at the heart of continuing professional education. Over the course of their careers, professionals change how they think, how they act in practice, and how they interact with clients. Professionals use their experiences as the basis for making these changes and for refining their practice.

Queeney (1996) suggested that continuing professional education, by definition, refers to the education of professional practitioners, regardless of their practice setting, that follows their preparatory curriculum and extends their learning throughout their careers. She goes on to say that ideally, this education enables practitioners to keep abreast of new knowledge, maintain and enhance their competence, progress from beginning to mature practitioner, advance their career through promotion and other job changes, and even move into different fields.

Bierema and Eraut (2004) suggested the term "continuing" in professional education assumes that there is some further development of initial training, a smooth transition (rare in many professions), and subsequent career progress. Jeris (2010) stated continuing professional education (CPE) refers to a particular area of interest within adult continuing education (ACE) that initially gained traction largely through Houle's work, beginning with articles he wrote in the 1960s.

In Houle's (1980) seminal study entitled *Continuing Learning in the Professions*, he referred to continuing professional education as lifelong study (p.7). He also said:

> No single course of action can resolve the difficulties in all these arenas (professions) of debate and conflict, but a pivotal need is for every professional to have the capacity to carry out his or her duties according to the highest possible standard of character and competence; one way to accomplish this is to require every practicing professional to embark on a path of lifelong study. (p. 7)

According to Kumar and Shah (2006):

> CPE is a life-long process through which individuals update the knowledge, skills and attitudes acquired during their education. It is usually self-initiated learning in which individuals assume responsibility for their own development. CPE is not an end itself, but it is a means to an end. (p.14).

Houle (1980) pointed out that "too few professionals continue to learn throughout their lives, and the opportunities provided to aid and encourage them to do so are far less abundant than they should be"

(p. 303). Even though CPE as a prerequisite for licensure had begun to take hold 10 to 15 years before he wrote his book, Houle could not have known that CPE would explode into what it currently is today. Authors such as Rockhill (1983), Darkenwald and Merriam (1982), and Lisman (1980) were calling for mandatory CPE, citing its importance to each profession and a step in the right direction, if not the cure for maintaining competency and proficiency.

CPE Arguments and Contested Issues: What are They?

The proliferation and increased knowledge surrounding CPE is not without its problems and detractors. As Jeris (2010) pointed out, "several scholars have expressed concerns about the need for lifelong learning" (p. 277) as opposed to education. Before Houle wrote his book, Illich (1977) and, after Houle wrote his book, Ohliger (1981) questioned whether CPE was sufficiently differentiated from education. Obviously, if there is not a distinction between the two, you cannot know what to expect as a favorable outcome.

Fast forward 20-years and Queeney (2000) suggested that "CPE is neither a guarantee of competence, nor the sole answer to competence assurance" (p. 375). In addition, there are problems with the way the terms professional and professionalism are defined. Specifically, the lack of consideration given to the social, political, and economic forces feeding into and flowing out of the professionalization journey of occupational groups minimized the impact of the definition (Tobias, as cited in Jeris, 2010, p.278). Moreover, Jeris and Armoacost (2002) suggested that CPE had not addressed cultural issues with CPE program and curricula.

Woolls (2006) suggested there are factors which make CPE inconvenient and disadvantageous to their participation. Three specific factors are travel, time, and cost.

Others include instruction that ignores the norms of behavior and communication, emotions of participants, and feelings of embarrassment.

Finally, Nowlen (1988) stated, CPE must go beyond providing information and teaching technical procedures; it must help professionals build their collaborative, judgmental, reflective, and integrative capabilities. It also must consider the individual practitioner's context, for "the relationship between continuing education and performance is unsatisfying when it is based simply on the relationship between a job description and an individual's knowledge and skills" (p. 69).

Informal Learning: What is it?

It has been stated that formal, informal, and incidental learning are components of learning. Therefore, I believe it is appropriate to define and take a close look at the definition of the broader construct – learning – before analyzing, comparing, contrasting, and discussing the sub-categories (components): formal, informal, and incidental learning. According to Marsick and Watkins (1990), learning can be defined as "the way in which individuals or groups acquire, interpret, reorganize, change, or assimilate a related cluster of information, skills, or feelings" (p. 4). They go on to say that learning "requires digging below the surface to identify and examine values and assumptions that govern the way situations are understood and addressed" (pp. 51-52).

Labelle (1982) and Mocker and Spear (1982) defined learning as university or college studies, short professional training courses, or externally planned programs of instruction.

Hrimech (2005) defined learning as learning sanctioned by an institution, such as a college or by a business, which leads to credits or some form of certification or diploma.

Hrimech, Labelle, and Mocker and Spear appear to be on the same page with their definition of learning. An example of a business that provides a certification is the Project Management Institute (PMI). PMI offers several certifications that typically require a person to have documented on-the-job-experience in the field of project management before sitting for one of their certification exam.

Informal learning, according to the National Center for Education Statistics (NCES), has gained significant traction. In 2005, over 70% of adults reported that they participated in informal adult learning activities (National Center for Education Statistics, 2007). Marsick and Watkins (1990) defined informal learning by stating the following:

> Informal learning, a category that includes incidental learning, may occur in institutions, but it is not typically classroom based or highly structured, and control of learning rests primarily in the hands of the learner. In addition, informal learning can be deliberately encouraged by an organization or it can take place despite an environment not highly conducive to learning. (p. 12)

Marsick and Watkins (1990) further defined informal learning as learning that takes place "in the normal course of daily events without a high degree of design or structure" (p. 14) and occurs "from and through experience" (p. 15). According to Hrimech (2005), informal learning is learning that is usually self-directed, independently pursued, and unregulated, often for the purpose of solving problems.

According to Marsick and Watkins (1990), there is a distinct difference between formal and informal learning. Formal learning is typically institutional-sponsored, classroom- based, and highly structured. Informal learning, a category that includes incidental learning, may occur in institutions, but it is not typically classroom-based or highly structured and control of learning rest primarily in the hands of the learner. Marsick and Watkins (1990) go on to define incidental learning as "a byproduct of some other activity, such as task accomplishment, interpersonal interaction, sensing the organizational culture, or trial-and-error experimentation" (p. 7). Incidental learning almost always takes place in everyday experience, often as a result of learning from mistakes and addressing the unexpected. In addition, incidental learning can substantially influence the development of professionals, and it typically takes place on a regular basis, even though people are not always conscious of it. Table 2 shows those items that characterize informal learning, formal learning and incidental learning for comparison purposes only.

Table 2

Characteristics of Formal Learning, Informal Learning and Incidental Learning

Formal Learning	Informal Learning	Incidental Learning
Instructor led	Self-directed learning	Learning from involvement
Classroom Based	Networking	Learning from Mistakes
Coaching & Mentoring	Coaching & Mentoring	Assumptions
Series of Planned events	Performance Planning	Beliefs
		Values
		Hidden Agendas
		The Actions of Others

(Garrick, 1998)

Cheetham and Chivers (2005) took the information provided by Garrick and reported specifically on informal learning. Specifically, Cheetham and Chivers' study and subsequent report included 20 professions and 700 professionals; they determined that there were ten commonly reported types of informal learning methods or experiences that helped professionals become fully competent experts. Shown below are the ten types of informal learning that help professionals become experts:

1. On-the-job learning
2. Working alongside more experienced colleagues
3. Working as part of a team
4. Self-analysis or reflection
5. Learning from clients/customers/patients
6. Networking with others doing a similar job

7. Learning through teaching/training others
8. Support from a mentor of some kind
9. Use of a role model (or role models)
10. Pre-entry experience

Cheetman and Chiver's asked the same 700 professionals, that provided the list of ten types of informal learning, to rank each of the ten types of informal learning in order of importance. The 700 professional came up with the ranking listed below:

1. On-the-job training
2. Experienced colleagues
3. Working as part of a team
4. Self-analysis or reflection
5. Learning from clients
6. Networking
7. Learning through teaching
8. Support from a mentor
9. Use of a role model (s)
10. Pre-entry experience

Informal Learning: Issues and Problems

The expansion and advancement of informal learning into the Internet, the virtual age, video games, simulations, and the digital age has not necessarily helped those that support informal learning gain credibility with supporters of formal learning. Many people still prefer the traditional, instructor-led format that characterizes formal learning.

Further, informal learning still has challenges and issues that may preclude it from supplanting formal learning as the learning method of choice. According to King (2010), the primary challenge of informal learning is the lack of universal access which leaves many unable to develop lifelong learning skills. Moreover, with lifelong learning becoming a necessity, minimized access could have a dramatic effect on those that are left outside, ultimately leading to disadvantages in the economic, professional, political, social, health, and personal domains (Brown & Thomas, 2006).

According to Marsick and Watkins (1990), informal learning and its subset, incidental learning, is unstructured in nature. Therefore, it is easy to become trapped by blind spots as it relates to one's own needs, assumptions, and values that influence the way people frame a situation, and by misperception about one's own responsibility when errors occur.

According to Hager (1998), there is a litany of problems associated with informal learning; specifically in the workplace. Hager (1998) goes on to say, Informal learning, "recognition of prior learning" is at once more easy and in some cases more difficult. It is easier in the case where overall recognition in a particular occupation is concerned.

For example, recognizing and supporting the work of a successful tailor without any formal training is easier to do if they have samples of their work, and testimonies from customers and other tailors. Consequently, on the basis of their work, they would be admitted to advanced courses on tailoring. However, without proof of their work, admission to an advanced program on the basis of knowledge and experience alone, becomes a more difficult proposition.

Davison's notion of equivalence as a criterion for making judgments about the educational significance of informal learning, but the end result is not objective. Davison's comments could be viewed this way: the process of determining equivalence and the significance of that equivalence is subjective. In many circles, subjectivity may not be the goal.

Schon (1983) suggested that informal learning does not provide the body of disciplinary knowledge that is required for the theory/practice component that a learner needs to build on the view, or substratum, that he calls "technical rationality." Without this disciplinary knowledge, which Schon believes is mostly scientific, learners do not have a basis for solving the problems that they encounter on a daily basis.

According to Hager (1998) there appears to be numerous variables that preclude researchers from isolating and highlighting the salient points of informal learning that actually accentuate the phenomenon, and the contributions it makes to learning.

Comparative Analysis: Project Management is a Relative Expertise Gained Through Deliberate Practice

I investigated four strands of literature: expertise, project management, CPE, and informal learning. The four strands were investigated in the order listed with expertise being a broad area representing many disciplines or professions. Experts and expertise can be found in all walks of life: cooks, seamstress, designers, tailors, painters, musicians, etc. I focused on one profession, project management. Project management is a discipline that has experts, CPE courses, informal training, and a certification process in place. The certification process is provided by the Project Management Institute (PMI). The PMI certification that is most relevant to this study is the Project Management Professional (PMP) certification.

The purpose of this study was to explore the project management domain – area of knowledge – in the EPCM industry and the expertise that underlies and delineates a project manager's competencies at the novice, intermediate, and expert levels for the purpose of understanding how these outcomes are achieved in the project management domain.

I used three studies in a qualitative approach to investigate project managers' growth from novice to expert in the project management domain. In order to effectively understand and explore project managers in the context of the EPCM industry, this study thoroughly investigated and researched the four strands of literature: expertise, project management, CPE, and informal learning that are integral components of the research question.

In the research on expertise, I discovered that there are two types of expertise: absolute and relative, and then there is a concept called deliberate practice.

Absolute expertise assumes greatness or creativity arises from chance and unique innate talent (Simonton, 1977). Therefore, absolute expertise is not necessarily developed and greatness and performance, by virtue of definition, are expected; as a result, absolute experts improve their performance with experience. In the absolute domain, there are several methods that can be used to identify an expert and expertise: retrospection, concurrent measurement, and independent index.

These methods (retrospection, concurrent measurement, and independent index) were briefly discussed earlier in this study, but were not elaborated on because, determining the definition of Absolute expertise was the most important research activity of this study, not the methods that can be used to measure Absolute expertise.

In addition, absolute expertise is not an expertise that can be experienced in the project management domain. Relative expertise, which has significance in the project management domain, assumes a separation and a comparison between those who are experts and those who are novice. In addition, the relative approach to expertise assumes that expertise is a level of proficiency that a novice can achieve. As a result, the relative domain has a less precise definition of expertise, because the comparison is made in a continuum with fewer boundaries.

Given the analysis above, where would/should project management fall and what should industries at large and companies in general expect in the way of expertise from project managers who work in the field of project management? Considering the nature of project management and the need for pragmatic experience, I would not expect to see any project managers fall into the absolute expertise domain. Specifically, I would not expect to see any 10-year-old project managers managing degreed engineers, managers, and administrative personnel with the responsibility of managing multimillion dollar projects. Therefore, that leaves the relative expertise domain, where there is a more applicable concept of a distinct separation between novice and experts, with an emphasis on the possibility of a novice achieving the status of expert over time.

However, given the nature of consultancy and the fact that most companies, including the one I worked for, try to bill out the hours of all consultants, managers, and engineers, there is very little time for training or CPE. As a result, a person may lament that it was and will be very difficult for a person in the relative domain to become an expert or reach an increased level of expertise. However, Ericsson (2005) suggested that our evolving and keener understanding of what distinguishes experts from novices should lead to better and more effective training (p. 235); which we hope leads to a higher level of expertise.

Given the information above, it can be stated that expertise and how it is developed can be found in the theory of deliberate practice. The goal of deliberate practice is to improve specific aspects of performance in such a way that attained changes or level of proficiency can be measured and integrated into a person's performance (Ericsson, 2006, p.698). The core assumption of deliberate practice is that expert performance requires the meticulous search for training and tasks that a person can master sequentially. The training can be found in CPE, and the task can come from project related activities.

The Project Management Institute (PMI), in the form of the PMP certification process, provides a framework similar to deliberate practice. PMI requires integration of task, training, and a four-hour, 200-question performance test as a prerequisite to PMP certification. PMP certification has lofty requirements. A person with a bachelor's degree must acquire 4,500 hours (two and a half years) of project management experience and 35 educational hours of formal project management training before sitting for a 4-hour, 200- question exam. In addition, he/she must score 61% or higher to pass the exam. A person without a bachelor's degree must acquire 7,500 hours (three and a half years) of project management experience, the same 35 educational hours required by a person with the bachelor's degree, and then pass the same 4-hour, 200-question exam with a score of 61% or higher.

In addition, there is an on-going component that requires a person who passes the PMP exam to acquire 60 professional development units (PDUs) every three years to maintain their certification. Basically, the last component requires all project managers to improve or upgrade their skills and knowledge every three years through CPE, teaching, writing books, publishing articles, etc., or they could risk losing their certification. PMI maintains a list of activities that can be completed with various PDU values. This ongoing requirement is a key step toward bridging the gap between expert and novice. However, it is not a cure-all. According to Nowlen (1988), CPE studies show that not all CPE training translates into improved job performance (p.69).

Summary

This literature review provided some interesting facts while uncovering and accentuating a void in the area that I believe justifies this study. *I believe it is important to communicate upfront that I believe Project management is a relative expertise that is gained through deliberate practice.*

Expertise is a broad strand of literature that can be categorized or broken into two general domains: absolute and relative expertise. The theory of deliberate practice can be used as a tool to develop expertise and essentially plays an important role in the journey professionals negotiate while moving from novice to expert.

Project management is a discipline in the broad area of expertise, and it can be defined as the application of knowledge, skills, tools, and techniques to project requirement to bring projects in on time, under budget, and within quality standards.

Project managers should be placed in the relative domain of expertise and skillfully guided to the pinnacle and status of expert project manager through deliberate practice and the utilization of PMP certification. PMP certification is a professional certification that requires years of experience, formal training, and completion of a four-hour 200-question exam.

CPE, or continuing professional education, and informal learning can be used by a project manager to enhance his/her knowledge, improve his/her expertise, and provide a lifelong learning agenda that, I believe, is essential to every project manager's continued growth and success. CPE can be defined as the education of professional practitioners, regardless of their practice setting, that follows their preparatory curriculum and extends their learning, throughout their careers. Informal learning is typically classified as self-directed learning, independently pursued, and un-regulated. It is usually associated with initiatives that require in-depth problem solving.

In project management, the Project Management Institute recently created two additional certifications that are considered higher certification than the project management professional (PMP) certification. They are called portfolio management professional (PfMP) and program management professional (PgMP). The same type of study carried out in this study for PMP can be initiated for expertise, CPE, and the portfolio or program management professional certifications. Finally, the ongoing questions regarding CPE and the impact or lack of impact it has on job performance can be investigated further.

As a result, there are still similar, if not larger, areas of research to study relative to this topic. However, the research and data gathered in this literature is the basis for the central research questions, conceptual framework, and proposed method of this study.

Chapter 3

Methodology

THE PURPOSE OF this study was to explore the project management domain — or area of knowledge — in the EPCM industry and the expertise that underlies and delineates a project manager's competencies at the novice, intermediate, and expert levels for the purpose of understanding how these outcomes are achieved in the project management domain. The results of this study could be used to help educators and researchers understand how expertise is obtained.

I used a qualitative approach in this study. A study is a bounded integrated system; a system that has settings, concepts, and samplings. In addition, it is believed to be an intensive description and analytical review of a phenomenon or social unit such as an individual, group, institution, or community (Stake, 1995; Merriam, 1998). Therefore, the most appropriate approach to achieve this study goal was to employ a qualitative study. A qualitative study allows for investigation of specific information about the values, opinions, and the social context of each project managers within each study, while the comparative approach enables the exploration of the phenomena that exist across the three studies in the context of the EPCM industry.

This chapter details the steps taken in the collection of data, the sampling technique(s), sourcing of participants, data analysis, peer reviews and member checks. This chapter closes with a discussion on the limitations of this study.

I selected one oil company and two well-known project management consulting firms in the EPCM industry for this study. The oil company is located in northwest Indiana and the Chicagoland area and has a project management department that is comprised of approximately 12-15 project managers. I interviewed a female project manager from this company, and she classifies herself as an intermediate-level project manager. The two consulting firms are located in the Chicagoland area, and they have project management departments that consist of 25-30 project managers. I interviewed one project manager from each company. One of them is a project manager with little or no formal project management training, but is considered by peers to be an accomplished project management expert. The other project manager is a PMI-certified PMP with formal project management training, but not necessarily an expert project manager.

Research Procedure

Data was collected through semi-structured interviews using an interview guide that includes interview questions. The interview questions are shown in Appendix A. Face-to-face interviews, 30 to 45 minutes long, were conducted at a location mutually agreed upon by the participants and me. In addition, participants were selected through convenient sampling and given an "Invitation to Participate form (Appendix B)". Table 3 shows a timeline of the interview process.

Table 3

Interview Process Timeline

Participant	Expertise level	Email Invite	Interview	Interview Transcribed	Member Check	Summarize findings
Brad	Expert	7/7/14	8/27/14	9/1/14	9/8/14	9/11/14
Karen	Intermediate	7/7/14	8/29/14	9/3/14	9/8/14	9/12/14
Jeffrey	Expert	7/7/14	8/30/14	9/5/14	9/8/14	9/13/14

Creswell (2005) defined convenient sampling as "a qualitative sampling procedure in which researchers intentionally select individuals and sites to learn or understand the central phenomenon" (p. 596). The fluid nature of this approach provides the greatest opportunity for the participants to share their experiences without regard to the limitations that are inherent in questionnaires, surveys, the Delphi technique, or other forms of qualitative data gathering. A transcriber was paid to transcribe the audio-recorded interviews, and I reviewed the transcriptions with my field notes to ascertain the themes to summarize my findings. In addition, I used structured questions to drive the interviewing process, while also using open-ended questions to capture the how and why of project managers' behavior in the context of the EPCM industry.

Creswell (2005) also suggested that in qualitative research, you ask open-ended questions to allow participants the opportunity to voice their experiences unconstrained by any perspectives of the researcher or past research findings. This method appears to be a suitable approach to capture the learning outcomes of a project manager at various levels along the novice-intermediate-expert continuum.

Before I began data collection, the study design was reviewed and approved by the Institutional Review Board of Northern Illinois

University. After the accounts of the subjects' learning experiences were explored, those experiences were analyzed using the research questions.

Participants

A convenient sampling approach was used to identify three project managers (PMs) for this study. I interviewed 3 PMs, their pseudo names are: Brad, Jeffery, and Karen, and their experience levels ranged from five to 30 years in the field of project management. These PMs were selected from a pool of project managers that I encountered over the past 25 years. Brad has little or no formal project management training, but considers himself an expert project manager. Karen, has formal project management training and considers herself to be an intermediate-level project manager. Jeffrey, is a PMI-certified PMP who has experience as a project manager and considers himself an expert on the novice-to-expert continuum.

Two of the PMs, Brad and Jeffrey, were chosen from separate for-profit consulting firms, and the third PM, Karen, was chosen from a for-profit oil refinery that employs consultants. It should be noted that in 2013- 14, I had some interaction with the PM that was chosen from the oil refinery. However, I do not have any work history with the other two consultant PMs. With that said, one of the PM consultants currently works for a company I interviewed with in the 2005-2009 timeframe. The PMs selected were initially contacted by phone and email, and those that agreed to participate were sent an "Invitation to Participate form (Appendix B)" and asked to respond within 3-5 days.

After receiving the forms and selecting the appropriate participants, I met with two of the three participants face-to-face. The third person could not meet me face-to-face, so that interview was conducted via a virtual tool called "Go-To-Meeting."

During the initial meetings I described the study, and the role the participants would play. I also shared the consent to participate form and the form that gave me the authority to audio-record the interview (Appendix C). According to Creswell (2005), it is typical in qualitative research to study a few individuals. Table 4 provides a demographic summary of the project managers who participated in this study.

Table 4

Participant Demographics

Participant	No Formal Project Management Education	Formal Project Management Education	PMI – PMP Certified	Novice (1-5yrs Exp.)	Mid-Career (5-15yrs Exp.)	Expert (15-30 yrs. Exp.)
Jeffrey		X	X		X	
Karen		X			X	
Brad	X					X

Instruments

As stated in the literature review section of this study, I endeavored to examine project managers in the context of the engineering procurement and construction management (EPCM) industry. In the process, a key component of the study was to determine the outcomes achieved by project managers as they move from novice to intermediate to expert project manager. In an effort to accomplish this goal, I developed interview questions that were aimed toward ascertaining the steps taken and the outcomes achieved by novice, intermediate, and expert project managers as they sought to achieve and maintain the level of expert project manager.

There are two sets of questions: one set for novice and one set for intermediate and expert project managers. The questions are designed to accentuate key experiences that led me to gather themes that help to facilitate a definitive commentary on the outcomes required to achieve the goal of expert project manager, as stated by the participants in the study. The interview questions are shown in Appendix A.

Data Analysis

According to (Merriam, 2002), data analysis should be conducted simultaneously with data collection. Using this approach allows the researcher to "test emerging concepts, themes, and categories against subsequent data" (p. 14).

During the review of the audio recordings, a paid transcriber transcribed the audio recordings. I then prepared reports for each participant and allowed each participant an opportunity to review their study and provide feedback. Next, I coded the information (transcribed information) from the recordings into five categories: project management, CPE, informal learning, formal learning, and an open theme before a final comparison across studies was completed. Figure 2 is an in-depth portrait or template of the three studies and how they were coded.

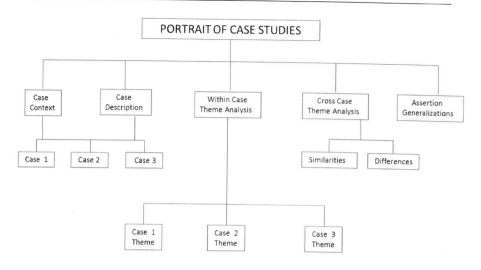

Figure 2. Portrait of three studies.

Figure 2, study 1, involved Brad; study 2 involved Karen; and study 3 involved Jeffrey. In addition, I discuss the context and description of each study, the themes in each study (project management, CPE, informal learning, and formal learning), the similarities and differences that exist across the three studies, and wrap up with a discussion about assertions and generalities.

Member Checks

Member checks were conducted on all interviews. Once the interviews were transcribed and I reviewed them, I sent electronic copies of the transcripts, via email, to the participant to ensure that the data collected from the transcripts was consistent with what the participant said. The process of sending participants copies of the interview transcripts ensured interview integrity and assured that I accurately captured the thoughts of the participants (Merriam, 2002).

Peer Review

For the purpose of inter-rater reliability, an independent research firm — a person not involved in the research -- served as a peer that partnered with me to analyze the data collected in the interviews as well as the themes identified from the transcripts of those interviews. The actual process was initiated and carried out through email and finalized through a follow-up phone call and conversation. This procedure allowed me to get unbiased feedback and provide someone other than me the opportunity to determine if my findings were consistent with the data collected during the interview (Merriam, 2002).

Limitations of the Study

There are several limitations to this study. First, this study investigated project management and project managers in the context of the EPCM industry. There are a number of industries that employ project managers and it would be reckless of me to assume, purport, or suggest that the results of my findings are a microcosm of all of the industries at large. Realistically, there could be a different dynamics in construction, banking, real estate, education, finance, telecom, human resources, and other industries.

Secondly, the sample size of three could have been larger. A larger sample size would have provided a more realistic understanding of project managers in the context of the EPCM industry. In the Chicagoland area alone, there are probably 2,000 to 3,000 EPCM project managers. This study provides limited information about the actual journey of a project manager from novice to expert.

Thirdly, this study investigated the journey of a project manager from novice to expert without regard to gender. A study could be done on the journey of a female project manager from novice to expert. By the same token, a study could be done on the journey of males from novice to expert.

The final significant limitation of this study is that it could have been done on a subset of the project managers, e.g., the journey of single females, single males, married females, and married males as they journey from novice to expert. I also could have selected project managers, male or female, married or single, in a certain age group, e.g., 25-35, 35-45, or 45-55. In summary, there are several permutations that could be investigated beyond this study

Chapter 4

RESULTS

This qualitative study involved an analysis of project managers in the context of the engineering procurement and construction management (EPCM) industry. Specifically, this research initiative used three studies to investigate how project managers become experts. The focus of the study centered on project managers who were at various levels of competency along the novice-to-expert continuum and sought to ascertain through interviews the answers to the following research questions:

1. How do project managers (PMs) in the EPCM industry conceptualize and define the role of an expert project manager?
2. How do project managers describe their evolution from novice to expert?
3. Are there any other factors that contribute to project managers becoming experts? If so, how are these factors related to the three types of expertise: absolute, relative, and the theory of deliberate practice?

Through my personal observation and inter-rater reliability, I came up with four major themes: (1) project management, (2) continuing professional education (CPE), (3) formal learning, (4) informal learning and one open theme, advice for people trying to break into project management. Figure 3 shows an updated picture of the initial conceptual framework shown in figure 1.

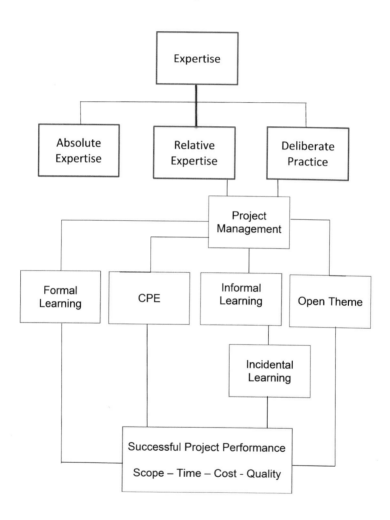

Figure 3. Conceptual framework updated (Walters, 2015).

A code-to-theme table, Table 5, shows how I came up with the themes from the transcribed interviews. It should be noted that incidental learning is a component of informal learning, so it will also be discussed with informal learning. Table 5 shows the code to theme for study 1 with Brad. I used keywords, sample quotes and phrases to help me determine the prevailing themes from the interviews. A complete table with information from studies 2 and 3 with Karen and Jeffrey respectively can be found in Appendix E.

Table 5
Code to Theme

Study 1 - Brad

Key Word(s)	Sample Quote	Phrase(s)	Theme
Project Management	Brad said he migrated toward project management	Brad did not have a project management mentor	Project Management
Prof Engineering License	Pursuit and maintenance of a PE license has a dramatic impact on his life	Brad has to maintain a CPE agenda by acquiring 30 PDUs every 3 years	Continuing Professional Education
Self-direction	I was a jack or all trades	I was responsible for everything	Informal Learning
Learned from Involvement	I watched another project manager	Brad learned from mistakes	Incidental Learning

Project management Professional (PMP)	I took a 1-day course in Project Management – 20 years ago	After I pass the PMP exam I will take more project management courses	Formal Learning
Advice	Find a mentor	Take a class, read books	Open Theme

As discussed in Chapter 3, I interviewed three project managers in the Chicagoland area; two of the project managers, males, work or have worked as project management consultants in the engineering procurement and construction management industry, and the third project manager, female, works for one of the oil refineries in the Chicagoland area. I used the format shown in Figure 2 to report the results uncovered in the interview for each study and then completed a comparative analysis across the three studies before closing the chapter with a discussion on assertions and generalizations. Table 4 provides an outline of the characteristics of the three project managers that were interviewed for this study.

The first project manager I interviewed, Brad, is a senior project manager who works for a consulting firm in the Chicagoland area. Brad has an undergraduate degree in chemical engineering and an MBA. However, until December of 2014, Brad had no formal project management training. He has worked 18 years as a process engineer or engineering manager doing project management work without the title "project manager." Brad has been with his current company the past 10 years for a total of 28 years in the industry. Typically, Brad's works includes pure project management, and he is in a position where he manages several project managers as they work with different clients in different industries to deliver a plethora of projects on time, under budget, and within quality standards. Using the information shown in Table 3, along with the way he learned his craft, the number of successful projects he has completed, the complex situations he has been exposed to, and his international project management experience, Brad considers himself an expert. Brad stated the following to support his claim:

> The progression from novice to expert took time. I spent 18 years on different projects including international projects, sharpening my craft without the benefit of being called a project manager. I've developed different formats, templates, and worksheets for tracking and reporting on projects over many project lifecycles and that process has helped me reach the level of expert. In addition, over the past ten years, I have successfully managed other project managers and been a conduit for knowledge transfer. Finally, I have successfully completed many complex projects over the past 28 years.

The interview took place in the library of a seminary in the Chicagoland area, and the following four themes emanated from the interview: project management, CPE, informal learning, and formal learning. In addition, there was one open themes that was discussed at the end of the interview.

Study 1

Project Management: How Project Managers Understand and Define Their PM Roles.

It is important to understand what project management is. Project management is the application of knowledge, skills, tools, and techniques to project requirements to deliver projects on time, under budget, and within quality standards; and it is accomplished through the appropriate application and integration of 49 logically grouped project management processes which are categorized into five process groups (PMI, 2018).

Little is known about how the majority of people get their start in project management. Some people migrate to project management because it seems interesting, other people find themselves thrust into project management and asked to complete project management initiatives, while others are given the title and told to figure it out.

Regardless of how people start, many times how they progress has a lot to do with how successful they are at delivering projects. During our interview, Brad characterized his start with a mixed bag of circumstances:

> Well, it was not so much a conscious decision as it was an evolution of necessity. I've worked in environments where there simply wasn't anyone else to drive a project so it became my responsibility. When I took some time to think about what to do next, continuing as a process engineer did not inspire me. It was more of a been-there-done-that, so I decided to transition into project management.

Brad went on to explain that project management is a leadership role that allows him an opportunity to cultivate his leadership skills while interacting with engineering disciplines. In addition, Brad stated, "The level of multitasking is generally quite high and often presents the biggest obstacle in project management."

When asked how he learned what a project manager does and how to do it, Brad stated the following:

> Well, I never had anyone mentoring me as far as project management goes. Nobody showed me the ropes or gave me pointers. I had supervisors or other people that I worked around that made suggestions, but typically, I would either observe what they were doing, or they would say, "Hey, why don't you try doing something like this," but there was never anything structured or formal as such. So I had to figure out a lot of things on my own. I used common sense or trial and error. Basically, I tried to emulate the things that worked, and tried not to do the things that didn't work.

Brad continued with an explanation on how he learned to solve problems that a project manager faces: "I observed what other people did and settled on what worked. A lot of problems can be solved if you think about them and use common sense, or what I like to call "learned common sense".

I asked Brad to explain what he meant by "learned common sense," and he replied: "In trial and error you learn over a period of time, and common sense comes after learning and falling on your face a couple of times."

Brad's commentary on learned common sense was a good transition into my next question. I asked Brad to describe the process of his growth from beginner, or novice project manager, to his present status, and he stated the following:

> Well, the progression definitely took time. I started as a process engineer and had experiences that, as indicated earlier, led me towards project management. So learning or knowing what should be done or needs to be done is not as difficult as learning how to execute the necessary tasks efficiently. It takes many project lifecycles and associated time to develop the systems that will be used to make project management efforts more efficient. In addition, it takes time to develop tools, good habits, and problem solving skills, whether the problems are people or technical problems.

It should be noted that Brad spent 18 years doing project management work, sometimes traveling internationally, before he was given the title project manager. He has since worked the past ten years as a project management consultant and senior project manager. I asked Brad to discuss the most important lesson he learned through his work as a project manager and senior project manager in the EPCM industry. Brad replied with the following:

> You can't just wing it [project management]! I have a project I'm working on right now that epitomizes the importance of attacking each project with the right tools, techniques, and mental approach. There's nothing real complex about the project at this point, but there are basic steps I still need to follow:
>
> 1. What is the timeline and scheduling for this project?
> 2. What procedures, tools, and good habits should I consider?
> 3. Then I must plan the work and execute the plan; you can't just wing it.

Time management, can be very difficult, especially if you're handling many projects.

Continuing Professional Education (CPE): Its Role in a PM's Evolution

According to Eraut (1994), continuing professional education (CPE) in the non-U.S. context is generally a category that resides within continuing professional development (CPD), and CPD is a broader categorization that encompasses formal and informal approaches to continuing learning. Conversely, in the U.S., the term CPE is more common and can be defined as the education of professional practitioners, regardless of their practice setting, that follows their preparatory curriculum and extends their learning throughout their careers. Typically, this education enables practitioners to keep abreast of new knowledge, maintain and enhance their competence, progress from beginning to mature practitioner, advance their career through promotion and other job changes, and even move into different fields (Queeney, 1996).

For the purpose of this study, I used the U.S. definition of CPE, and in my interview with Brad, I asked him what events or activities had the greatest impact on his professional development as a project manager. Brad gave me two examples, and one of them can be classified as a CPE event. Therefore, I will discuss that event here. The other event will be discussed in the informal learning section that follows this CPE section.

Brad stated that going through the process of becoming a licensed professional engineer (PE) had a dramatic positive impact on his professional development as a project manager. Brad went on to discuss the following:

> To become a licensed PE, in my case for chemical engineering, requires that you pass a two-part exam. The first part of the PE exam involves basic engineering questions about all of the engineering disciplines which are covered in an undergraduate engineering program in your first two years of college.
>
> The second part is specific to your chosen discipline, in my case, chemical engineering. But preparing for that test and ultimately passing it was very educational. It gave me the ability to understand what the other disciplines (electrical, mechanical, civil, structural, etc.) do, and what was important to them.

It should be noted that Brad is an engineering project manager and typically his projects include all of the engineering disciplines: chemical, electrical, mechanical, and civil, along with other professionals like draftsmen.

Therefore, acquisition of a PE license could play an important role in successfully managing projects that include engineering professionals. In addition, Brad is required to perpetuate a CPE agenda to maintain his PE license. He must accumulate 30 PDUs every three years to maintain his PE license. However, with that said, Brad is a senior project manager. Therefore, I asked him if he felt he should be pursuing project management courses as part of his CPE agenda, and he said that it is a possibility after he passes the PMP (project management professional) certification exam. Brad recently attended a PMP prep course because he felt the need to top off his career with PMI's PMP certification. In his words, "I wanted to find out what I don't know."

A company's work environment and what they advocate in terms of continuing professional education can have a dramatic impact on their employee's ability to seek out, secure, and perpetuate a CPE agenda that leads to lifelong learning. As mentioned earlier in the literature review, Houle (1980) suggested that "too few professionals continue learning throughout their lives, and the opportunities provided to aid and encourage them to do so are far and less frequent than they should be" (p. 303).

Brad and I discussed the current climate in his company and whether or not that climate has influenced his professional development. He replied with the following statement:

> Well, I'm in an environment where I can observe how a lot of people similar to me do their work and that's very beneficial, but at the same time, we're all project managers, we're all working on our own projects, so we're not generally interacting directly with each other; it's more in passing.
>
> So in that respect, there's not a lot of formal interaction, but there's a very cooperative atmosphere in the company. In addition, I'm doing more pure project management. But one of the more positive aspects of my job is the increased activity and formality surrounding our use of project controls. The enhancements with project controls have come from my supervisor, who is a PMP.

Brad mentioned the words "pure project management" and "project controls." When he said he was doing pure project management, he meant he was using the tools and techniques of the project management profession, and his job responsibilities were related to project management only. Brad also mentioned project controls; basically, the use of project controls means he is using the tools and techniques that are required to initiate, monitor, and track a project's budget and schedule.

Brad closed the CPE section of the interview with the following thoughts on whether or not there was anything outside of his professional life that affected his development as a project manager:

> Not really! There may have been a few miscellaneous classes that I took in the past. But there's nothing outside of the working world. There really hasn't been that many situations that I can think of that would apply.

Informal Learning: Its Role in a PM's Evolution

As discussed in the literature, informal learning is not classroom based but rather takes place "in the normal course of daily events without a high degree of design or structure" (Marsick & Watkins, 1990). In addition, we learned that incidental learning almost always takes place in everyday experience, often as a result of learning from mistakes and addressing the unexpected. Moreover, incidental learning can substantially influence the development of professionals, and it typically takes place on a regular basis, even though people are not always conscious of it (Marsick & Watkins, 1990).

During my interview with Brad, I asked him about events that had a dramatic impact on his professional development, and in the last section on CPE, I indicated that one of the two events Brad mentioned would be covered under informal learning. The following informal learning event includes some of the elements of informal learning shown in Table 2. Brad shared the following informal learning event:

> A company that I worked for decided to do a joint venture project in Shanghai with two other partners. The first partner was one of our customers, and the second partner was a Chinese land developer. At the

time, I was an engineering manager for my company, and because it was a very small company, I functioned as a jack-of-all-trades; anything and everything kind of came through me. What I really liked about the project was I was there from day one. I was responsible for the process design, most of the project management, directing different engineering folks, as well as doing the engineering myself. I was also in charge of construction, training, and startups.

I asked Brad to summarize the informal learning components, based on Table 2, which he felt were most evident, and he said, self-direction, performance planning, and networking. He also said the international travel was very educational. Moreover, during the interview, Brad cited lesser situations in his professional life that fell in the category of informal learning. They were on-the-job learning, self-analysis or reflection, and use of a role model.

Incidental Learning – A Component of Informal Learning: Its Role in a PM's Evolution

Incidental learning, according to Marsick and Watkins (1990), is a subset of informal learning and is unstructured in nature. In addition, Garrick (1998) stated that incidental learning takes place in the following scenarios: learning from involvement, learning from mistakes, contemplating assumptions, the actions of others, and through personal beliefs. When I asked Brad about people who had the greatest impact on his professional development, he cited a project manager who he learned a great deal from when he worked as a plant engineer. The situation and event he explained can be categorized as an incidental learning event based on Garrick's (1998) book. Brad shared the following:

> When I was a plant engineer in an ethylene plant, I worked very closely with a project manager – going back 22 years ago – and that gentleman is still there; he was just a force of nature. He was amazing at what he could accomplish and how focused he was on what was important, and he didn't worry about what was not important. And just working with him, watching the way he worked was very impressive and, I tried to emulate him in many ways. He was just very efficient.

Brad's example is indicative of incidental learning because he learned from involvement, the actions of others, and from mistakes. Brad went on to explain that he has learned from superiors, coworkers, and subordinates:

> I guess I'm always observing other people around me, including the people that report to me. I'll try to learn from them just like they try to learn from me. Generally speaking, at my company, there are a lot of other PMs, and although they might have different styles, there's always something to learn from them. I've been with my current company for ten years now, but only lately have I had an opportunity to observe many people with diverse backgrounds at the same time and learn from all of them.

Finally, Brad placed a lot of importance on his ability to learn from others by observing their actions and being involved in what he is managing.

Formal Learning: Its Role in a PM's Evolution

According to Marsick and Watkins (1990), formal learning is typically institutionally sponsored, classroom based, and highly structured. During his career, Brad has not had a lot of opportunities to take advantage of formal learning. The majority of his career has been crafted with informal and incidental learning scenarios. I asked Brad if he attended any formal project management courses, seminars, or webinars that could have enhanced or impacted his career. Brad indicated, "I took probably a one-day project management course 20 years ago, maybe." Brad went on to explain the following:

> I was not a project manager per se when I took the one-day project management course, but I was doing project management work, so to some extent I'm sure I learned something from the course, but nothing real extensive. As I mentioned earlier, I am a licensed professional engineer (PE) and that requires me to pursue continuing education to maintain my license. I try to keep it interesting, but I'm running out of topics. Typically, I will try to pick up a few hours every couple of years that are project management related.

I followed Brad's comments with the following statement and question: Now, that's interesting that you take formal education courses to maintain your PE certification, and you work as a senior project manager. Do you feel deficient or slighted by not regularly taking project management courses that could potentially make you a better project manager? Brad replied with the following:

> I guess after I take the PMP exam and pass it, I will consider taking more project management related courses. I feel it is part of my continuous improvement. I got my chemical engineering degree in 1986, I got my PE in 1996, I got my MBA in 2006, and then I started wondering what I was going to do for 2016, so I am just a little bit ahead of the curve. Every once in a while I

get this urge that there's something I don't know that I should probably find out. So, continuing in project management and becoming a little more specialized or just better at project management is my immediate goal.

Recently, Brad attended a four-day PMP prep course that was facilitated by Walters Consulting, LLC. Basically, Brad felt the need to determine, by attending the course, what he didn't know and, in the process, prepare for the PMP exam.

<u>Open Theme – Brad's Advice for Those Getting into Project Management</u>

I asked Brad to take a moment, review his career, and then answer the following question: What advice would he give to people trying to get into the field of project management? Brad suggested the following:

Okay, I mentioned earlier I did not have a mentor, and I didn't take any project management courses. I picked up a few Internet courses, in addition to the one I mentioned earlier, but for those that want to get into project management I would recommend they do the following:

1. Get some form of formal education, be it a semester course or a series of courses, classes, or seminars.
2. Find a mentor or a company that has mentoring. Again, mentoring is something that's always slipped away from me or hasn't been available to me.
3. Learn and understand how the engineering disciplines work.
4. Initiate and cultivate a continuing education agenda. If you cannot attend classes, then read the appropriate books.

5. Finally, make a conscious effort to practice the tools and skills they learn. Most of the time there will not be opportunities to use 100% of the new tools and skills you learn.

Brad went on to say that those who are in project management but find themselves at the novice end of the project management novice-to-expert continuum should do the following:

1. I think novice should do the same thing as those individuals trying to get into project management.
2. Learn what being a project manager means, and the best way to do that is by observing what project managers do.
3. Read books, secure a mentor, practice, and work hard. When you're reading the book, try to practice it.

I closed my interview with Brad by asking him what he does to maintain or sharpen his craft. Brad replied with the following explanation:

1. Well, I maintain my PE license because I am still a chemical engineer, and the PE requires that I perpetuate a continuing education agenda.
2. I don't allow myself to get stagnant. I believe you have to keep learning new things or refreshing the things that you have learned.

After his initial response, I asked Brad to elaborate on the project management component of his craft. Specifically, what does he do to maintain or sharpen the project management component of his craft? Brad replied with the following statement:

It's the tools. Typically, I try to solve problems by thinking about the tools that will help me solve the problem as easy as possible today, and possibly ten more times in the future.

For example, while attending your PMP prep course (Walters Consulting, LLC), I often spent time contemplating how I could organize different aspects of the material, simplify it, and make it easier to recall. As a result, I determined that there are a lot of charts and graphs that I can use to help me accomplish my goals.

In summary, the interview and study with Brad can be characterized as informative, unique, and eye-opening. Brad has an undergraduate degree in chemical engineering, and an MBA. He ascended to his level of expertise without the benefit of formal training, a mentor, or coach. He performed project management work for 18 years without the title of project manager, and he learned project management through, what he calls, the "school of hard knocks" and "learned commonsense".

In addition, during our interview, Brad was studying for the PMP exam. Shortly after our interview, 6 months, Brad passed the PMP exam, and he is now a PMI certified PMP.

Study 2

This study continued with a qualitative analysis of a female project manager in the context of the engineering procurement and construction management industry (EPCM). Specifically, the study investigates how project managers, regardless of gender, become experts. The focus of the study centered on project managers who were at various levels of competency along the novice-to-expert continuum and sought to ascertain through interviews and transcripts the answers to the following research questions:

1. How do project managers (PMs) in the EPCM industry conceptualize and define the role of an expert project manager?
2. How do project managers describe their evolution from novice to expert?
3. Are there any other factors that contribute to project managers becoming experts? If so, how are these factors related to the three types of expertise: absolute, relative, and the theory of deliberate practice?

The interview questions used in the first interview were also used in this interview; therefore the four themes that surfaced in the first interview were also apparent here. More importantly, using the same themes helped simplify the comparative analysis across the three studies. For the sake of clarity, the themes in this study are: project management, continuing professional education (CPE), formal and informal learning, and one open theme, advice for people trying to get into project management. Again, it is important to note that incidental learning is a component of informal learning, so it will be discussed with informal learning.

As I did with the first study, to maintain consistency, I used the format shown in Figure 2 to report the results uncovered in the interview for this study and then completed a comparative analysis across the three studies in this chapter before closing the chapter with a discussion on assertions/generalizations.

The second project manager that I interviewed, Karen, is a senior project manager who works for an oil refinery in the Chicagoland area. Karen has an undergraduate degree in civil engineering and an MBA. She has participated in company specific courses related to project management, construction management, cost control, negotiations, and safety. But she does not feel those courses should be classified as formal project management training. In fact, when I asked Karen if she had attended formal project management training, she said, "no." In addition, Karen is not currently pursuing PMI's PMP certification, but she acknowledged that it is an important certification and something she would like to pursue in the near future. She has worked for the same oil refinery for 14 years and characterizes herself as a fairly experienced, intermediate level project manager.

Karen is in a unique situation because approximately 99% of the time she is considered the "buyer" in the buyer/seller project scenarios that play out in the EPCM industry. To put that into proper perspective, Brad, the first person I interviewed, is usually referred to as the "seller" and would typically report to Karen if he was assigned/chosen to initiate and complete project work in the oil refinery that Karen works in.

Karen does project work, but a big percentage of her job requires her to assist people like Brad when they have refinery-related questions. Like Brad, when Karen is managing projects, she is responsible for managing several engineering professionals, draftsmen, and designers. The interview with Karen took place at a library in the Chicagoland area, and as I mentioned earlier, the following four themes surfaced from the interview: project management, CPE, informal learning and formal learning; there was also one open theme, advice for people trying to break into project management.

Project Management: How Project Managers Understand and Define Their PM Roles

Project management is proliferating, and companies in all industries are adopting the project management profession and the methodologies associated with it. As a result, many people are transitioning into the profession. However, it is still unclear how many people get their start, and once they get their start, how they improve their ability to perform and deliver projects successfully and at a highly efficient rate or level. More importantly, how do they maintain their ability and consistently deliver projects at a high level? Karen's start in project management was a little different than Brad's (study 1), in that she was recruited into project management after completing her undergraduate studies in civil engineering. Karen shared the following:

> I studied civil engineering in undergrad and I guess I assumed upon graduation that I would be working for a company as a civil engineer. But, while working in internships and coops, I knew that civil engineering wasn't the right fit for me.
>
> So, when the company I currently work for recruited at my school, I was disappointed that they were interested in me as a civil engineer.
>
> However, as an alternative, they discussed with me their project management workload, and it sounded interesting. So I applied for the job, got accepted, and began work in their project management department; I have been on this journey for 14 years now.

When I asked Karen about her early years as a project manager and how she learned what a project manager does and how to do it, she replied with the following:

> In my case, it was mostly hands-on learning and learning from other project managers. In addition, there was coursework that functioned as a guide to communicate concepts like: "This is how we manage projects here. This is how you do cost control. This is

> how you do construction management, and this is how you do project management." Summarily, I would sit in a class for a week or two, learn a little bit, and then go out and implement what I learned in class. In the process, I would learn from other project managers.

Karen went on to explain that formal learning, informal learning, continuing professional education (CPE), asking questions, and practical experience also played an important role in her learning process, as it relates to what project managers do, how they do it, and how they solve problems. Karen shared the following:

> I would say all three of the items you mentioned: formal learning, informal learning, and CPE were very important to my understanding of the project management profession. But in my case, my learning process began with formal learning, and I think that the education I got, even as a civil engineer helped me in my early progression. The engineering disciplines teach you how to think, and so that was the basis of the training I think I needed to succeed the way I have. So my career started with the formal learning, and CPE has kind of happened throughout my career. But formal learning was a big part of my training in the beginning. In addition, I spent a lot of time asking questions … I learned how to solve project management problems by asking a lot of questions.

Karen continued by saying the informal training she received early in her career was the most beneficial of any training she received.

I subsequently ask Karen about her progression from a beginning (novice) project manager early in her career to her status now, as an experienced project manager. Karen stated the following:

Well, the company that I work for prefers to grow their employees from the ground up and get them acclimated to their systems early on.

1. So I started out doing cost control, working under a cost controls manager who in turn was working with a project manager.
2. While doing cost control, I worked with a whole team of people who had various roles and responsibilities with a project.
3. I was assigned coursework on certain aspects of a project, like estimating, and then assigned to a field project where I got to see how estimating was done.
4. While in the field I was required to attend project team meetings and report on the aspects of the project that I was assigned.

Finally, as part of Karen's growth, she said she spent a lot of time asking questions. Karen said she is the type of person who would rather ask a colleague how they solved a complex problem than spend a lot of time trying to solve it alone. She also considered this aspect of her personality a very important component of her growth from novice to expert.

It should be noted that Karen is explaining a planned and structured approach to her development, an approach that was systematically used and applied in her growth from novice project manager to her present status. This approach was very different from the "school of hard knocks" that Brad experienced and endured during his progression from novice to his current status.

I asked Karen to share the most important thing she learned through her work as a project manager. Karen stated the following: "I would say the most important thing that I've learned, and I've learned a lot, is":

1. How to be a project manager.
2. How to better understand the significance of roles and relationships in a project and how they could dramatically impact the outcome of a project.
3. Confidence, you must acquire confidence. If you don't gain and show confidence, don't expect anyone else to show confidence in you.

<u>Continuing Professional Education (CPE):</u>
<u>Its Role in a PM's Evolution</u>

According to Queeney (2000), for CPE to have an effect on professional competence, it must address practitioner's educational needs or areas of weakness in the workplace (p. 376).

In my interview with Karen, I asked Karen what events or activities had the greatest impact on her professional development as a project manager. Karen cited two examples, but I will discuss one of the examples as a CPE event, while the other one will be discussed in the informal learning sections of this study.

> The first example is associated with my first assignment as a project engineer which in my case was one step before becoming a project manager. So I had attended classes, internally, on cost, completed all the project cost stuff, and then moved into the role of project engineer. Unfortunately, I didn't have a lot of guidance, and it didn't go that well. I was sort of thrown into that situation. It was different than anything I'd ever done. It was unlike anything anybody had ever really done, and I was working with a mentor.
>
> It was at this point I started learning how roles and relationships can impact your project. Unfortunately,

the person they assigned to mentor me became very ill and passed away.

I interrupted Karen and inquired about the stress and emotional strain she must have felt dealing with death and the prospect of continuing a project that she felt ill-prepared to complete. She continued:

> Yes! That was a bad situation in general. I never had the benefit of being mentored and neither had I, up to that point, cultivated a relationship with anyone who could actually help me. So I felt like it wasn't a very successful situation for me, because I think I was expected to do more than I was capable of at the time. And, that situation taught me a lot about expressing myself, and not being afraid to speak up and say, "Hey, I can't do this," or "Hey, I need a little bit more help with this than I currently have."

I asked Karen, if this situation impacted her confidence. She continued:

> Yes! Absolutely, and it made me question whether or not I was cut out for this [project management]; but luckily I stuck with it. However, that was number one and the first step along the path that made me start to evaluate my confidence, the role that relationships play, and the importance of talking with people and getting that guidance and support you need when required.

As I mentioned in Brad's study, a company's work environment and what they advocate in terms of continuing professional education (CPE) can have an impact on the lifelong learning initiative of a project manager. It can be argued that lifelong learning should be an integral component of every project manager's professional development agenda.

Karen suggested in our interview session that CPE was encouraged during her journey along the novice-to-expert continuum, but recently it had been abandoned because of what she characterized as a shift in management philosophy. I asked her about the current atmosphere in her department at work and she provided the following comments:

> Currently, the atmosphere in the department is very negative, and that negativity has come from the top down [from upper management]. I think in any profession, especially project management, there must be a positive environment. If there is negativity at the top, it will permeate the department and have a negative impact on morale and work output in general. Unfortunately, it has become very apparent that our group has become very systems focused as opposed to people focused.
>
> When you're managing a project, I agree, you should adhere to the system and you should use the tools and follow the rules, but you can't lose sight of your most important resource, people.

Based on Karen's comments about her current work environment, I asked her how the negative environment has impacted her desire to improve as a project manager. She communicated the following:

> It's very hard. Because on the one hand, I like project management a lot. I like doing it, and I like the people that I'm working with. But it's very hard to want to improve when you don't feel, number one that your improvements are being recognized, and, number two, that those improvements will get you anywhere.

The most important thing to note here is that Karen works in an environment unlike the one that Brad works in. Karen feels stunted in her ability to grow and doesn't feel supported by upper management. As a result, she appears to be struggling with doing well, and the prospect of doing well while not being recognized for it is a conflict that represents a dichotomy that she does not want to deal with. Karen pointed out that a project manager should be put in a situation that supports and cultivates his/her desire to become an expert, or the possibility of becoming expert is severely diminished.

Informal Learning: Its Role in a PM's Evolution

Earlier in this study, I determined that informal learning is a form of learning that can occur in a classroom setting but is not usually classroom based. Typically, this form of learning takes place during the normal course or progression of daily activities and is primarily under the control of the learner. Examples of informal learning would be self-directed learning, networking, and mentoring. In addition, incidental learning, a subset of informal learning, can influence the development of professionals, and it usually takes place regularly. Examples of incidental learning are learning from involvement, learning from mistakes, and learning from assumptions.

In my interview with Karen, I asked her what events or activities had the greatest impact on her professional development as a project manager. I discussed one of her two examples in the last section, CPE, and I will discuss her second example in this section. The example is really two examples in one. Karen compared and contrasted her experiences during a plant shutdown when she was a project engineer with her experiences in a shutdown as a project manager years later in her career. Karen commented:

There were two situations that stand out in my mind as events that impacted me. Over the years, I have been fortunate enough to work two different shutdowns in two roles:

1. The first shutdown, I was a project engineer and it served as a confidence builder because I had a small amount of mentoring, and I managed, as a project engineer, to get a lot of work done. More importantly, I think I gained more experience in the 30-day shutdown than I did the whole year prior working as a project engineer.

2. In a second shutdown, I was a project manager, and I was kind of thrown into the shutdown, but this time, as a project manager and because I had worked a shutdown earlier in my career, I was able to perform at a high level and get a lot of work done. But those two situations had a dramatic effect on my development.

It should be noted that Karen's examples include two forms of learning from Table 2. She mentioned that she was mentored as a project engineer, and she stated that the next time she worked a shutdown as a project manager, she was better prepared to deal with the responsibilities associated with it. In Karen's words, "I am a big proponent of learning by doing." Learning by doing is congruent with incidental learning, which means learning by involvement.

Incidental Learning – A Component of Informal Learning: Its Role in a PM's Evolution

Marsick and Watkins (1990) created an informal learning model that integrated incidental learning. They ascertained that incidental learning is always occurring, with or without our conscious awareness, and that incidental learning can be triggered by an internal or external stimulus (p. 29). I asked Karen to discuss the people who had the greatest impact on her professional development as a project manager. She cited two examples of incidental learning; the first took place in a classroom while she was completing her undergraduate degree in civil engineering, and the second took place at her place of employment. Both examples can be categorized as incidental learning events under networking, learning from mistakes, and learning from involvement. Karen communicated the following:

> Becoming a mom, hands-down, has had an impact on my time management skills; especially as a full-time working mom. I have great support at home, and my husband's fantastic. But even though we share the responsibilities, there is still a lot to do, including spending quality time with the kids. When it comes down to it, if I did not have time management skills, I would definitely go under.

Karen's first example, by definition, is a prime example of learning from involvement and the actions of others. Time management is one of the triple constraints of scope, time, and cost. It is definitely something to be cognizant of and look to improve on. It should be noted that poor time management can have a dramatic negative impact on your project and cause your project to be delivered outside of your client's requirements if ignored.

Karen continued with her second example:

> The second example involves my colleagues, specifically, other project managers. The way our organization is structured, we all sit together, and it makes it easy to

communicate. If I have a question, I just walk next door or across the hall and ask my question. And then there are people in the construction industry who I consider my colleagues. I cannot overemphasize the importance of developing relationships with contractors and suppliers and understanding how they work and do their jobs, because you need them. You can't sit in a bubble and declare that, "I'm the project manager and you're my contractor; do my bidding." It's a work relationship, and to be effective you must be aware of that.

Karen closed our discussion on informal learning with her thoughts on informal learning and the impact and/or non-impact of the mentoring process. Karen said she was mentored early in her career for about one year, and after that she had very little mentoring.

But, she also commented that very few of her colleagues have been mentored. The majority of her learning process, other than the formal training early on and the mentor she had as a project engineer 13 years ago, has taken place through involvement, the actions of others, and learning from mistakes.

Formal Learning: Its Role in a PM's Evolution

Learning sanctioned by an institution such as a college or business that leads to credits, certification, or a diploma is called formal learning (Hrimech, 2005; Merriam, Caffarella, & Baumgartner, 2007). Typically, this type of learning includes an instructor, curriculum, and an evaluation process. In addition, as shown in this (second) study, formal learning can be defined as institutionally sponsored or classroom learning, and is usually highly structured. During her career, Karen took some formal education courses, which can be classified as formal learning, and by her own observation, they have been very helpful.

However, I asked Karen to share those items or situations that were outside of her professional life that have or could have had an impact on her development as a project manager. Karen shared the following:

> The first example involves a professor I had. I remember him as a passionate professor because of the way he taught us, but it wasn't until later in life that I was able to put what he taught us into proper perspective. He was a construction management professor, and construction management was a new field of study within civil engineering at my school. He tried to teach us the way it was in the real world because he was from the industry. At the time, we were too immature to appreciate the importance of what he taught us. But now that I look back on it, I realize that what he taught us was really good stuff and relatively life changing. So I'd say he was one important influence.

Karen went on to discuss another formal learning event, completion of an MBA, and how it did not provide the dividends she expected or wanted from her current company. She had hoped the company she worked for would promote her to a new position.

> I spent three years getting my MBA from one of the top business grad schools in the country, and I was fortunate that my company paid for it. But if I had paid for it, I think I would've marketed it and went into a completely different career.

> I knew going in that it wasn't going to benefit me as a project manager, but knowing that and then facing the reality that nothing really changed is still somewhat

frustrating. [In summary], although the MBA has not resulted in a promotion, it has benefited me other ways in my professional career.

Karen and I discussed the importance and potential impact of the project management professional (PMP) certification, and she agreed that it is something she plans to pursue in the near future. Karen had the following comments about the PMP certification process:

> I'm not just saying this because I know you (Walters Consulting, LLC) teach the PMP prep course, but I have really been considering PMP certification. It's not something I need in my current role, and it's not going to get me much were I am ... but it's always good to have that information, knowledge, and the certification that goes with it.

During our interview, I suggested that Karen plan to attend my PMP prep course so she could begin the PMP certification journey, much like Brad did.

Open Theme – Karen's Advice for Those Getting into Project Management

I asked Karen to take a moment, review her career, and then answer the following question: What advice would she give to people trying to break into the field of project management? Karen suggested the following, "I would say it is important for a person to consider the following:"

1. Get formal training beforehand.
2. Spend time understanding what project management is all about.

3. Get a PMP certification.
4. Consider taking non-credited project management courses if you don't attend formal training.

Karen went on to say that those that are already in the field of project management but find themselves at the novice end of the project management novice-to-expert continuum should do the following:

1. I'd recommend getting as much experience as possible.
2. Find out what tools are available to you and practice using them wherever possible.
3. Finally, it is important to build relationships. As a novice project manager, you need a network or support system that you can draw from.

I closed my interview with Karen by asking her what she did to maintain or sharpen her craft. Karen replied:

1. In my case, I'm currently looking to work on larger projects. I think that's one area in my development that will go a long way toward improving my project management acumen.
2. In addition, I seek out training opportunities that I think will benefit me, i.e., the PMP certification.

    Moreover, I have enrolled in my company's culture of health program. Along with project management tools, I think it is important that I initiate and perpetuate a wellness program that supports a healthy lifestyle.

3. I'm also considering online courses that focus on organization and time management skills, although I have improved, I can still get better in these areas.

In summary, the interview and study involving Karen was a lesson in opposites, as it relates to Brad's study. Karen has an undergrad in civil engineering and an MBA. She ascended to her level of expertise through limited mentorship, formal training, and structured learning events. However, later in her career Karen's evolution was crafted through involvement, the actions of others, and learning from mistakes. Karen believes in the importance of PMP certification, but has not, to this date, January 1st, 2017, passed the PMP exam.

Study 3

In the third and final study of my research, I interviewed a male, Jeffrey. Jeffrey currently holds the position of president at an engineering procurement and construction management (EPCM) company in central Illinois. The focus of this study is concerned primarily with project managers who were at various levels of competency along the novice-to-expert continuum and sought to understand through interviews, and transcripts the answers to the following research questions:

1. How do project managers (PMs) in the EPCM industry conceptualize and define the role of an expert project manager?
2. How do project managers describe their evolution from novice to expert?
3. Are there any other factors that contribute to project managers becoming experts? If so, how are these factors related to the three types of expertise: absolute, relative, and the theory of deliberate practice?

The interview questions used in the two previous studies are used here and, for the sake of continuity, the same themes are used as well: project management, continuing professional education (CPE), formal and informal learning and one open theme, advice for people trying to get into project management. Incidental learning, a subset of informal learning, is covered in the discussion on informal learning.

Jeffrey oversees all training and hiring initiatives, including those that involve project managers. He also assumes some responsibility for business development. Typically, Jeffrey provides project managers to companies via a consultancy model, meaning companies contact Jeffrey and ask him to provide qualified project managers to work on their company site for short- and/or long-term assignments. The assignments can range from one to three years in duration.

Jeffrey has an undergraduate degree in mechanical engineering, and he is a PMI- certified PMP. Prior to becoming president, Jeffrey was a vice president, and prior to that a project manager, and before that, he spent some time doing engineering work. He has about 15 years' experience as a project manager, and he has taught project management courses at university and company sites throughout the Midwest. Typically, Jeffrey, like Brad, is considered the "seller" in the project scenarios that play out in the EPCM industry. The interview with Jeffrey took place via Go-To-Meeting, an Internet-based application that allows two or more people who are geographically displaced to meet and discuss business or other topics. We chose this format because Jeffrey was under the doctor's care and confined to his home. Moreover, the geographic distance between us, coupled with my need to record the interview, necessitated that I seek a practical and economical method like Go-To-Meeting.

The themes that follow are the same themes that were discussed in my interview with Brad and Karen and they are intended to explore and uncover the same or similar information uncovered in the interviews with Brad and Karen.

Project Management: How Project Managers Understand and Define Their PM Roles

Project management transcends industries, and many companies across the United States and the world have adopted project management as a means of achieving their goals and objectives. However, with the proliferation of project management, increased company awareness, and the growing trend of people transitioning to the profession comes the burden of uncertainty and questions. People want to know how project managers get their start, how PMs grow from novice to expert, and what PMs do to maintain their expertise.

Jeffrey, like Brad and Karen, began his career in a manufacturing environment. He started doing engineering work, which eventually led him to college and a career in project management. However, unlike Brad and Karen, Jeffrey was doing engineering work prior to completing his undergrad degree in engineering. In fact, the work that Jeffrey was doing caused him to pursue an engineering degree. Jeffrey shared the following thoughts on his start in project management:

> Well, I was working in manufacturing and started working myself into an engineering role, which led me to attend college. I attended the University of Southern Illinois University (SIU), Edwardsville, and obtained a degree in mechanical engineering.
>
> During my time there (SIU), I found myself managing a lot of projects. So a few of my classmates and I decided to enter several design competitions sponsored by Boeing, Ford, and other companies. The competitions took a lot of organization, and that responsibility usually fell to me.

Once I graduated, I was hired by a project management company who needed an engineer. I began my career in a dual role doing engineering and project management work after three months with the company. It was then that I realized how project management could help engineers get work done more efficiently. As a result, I focused more on project management and less on engineering.

I asked Jeffrey how he learned what a project manager did and how to do it. Jeffrey said he learned what a project manager did through company (internal) training and interaction with customers whose expectations helped him craft a definition of what a project manager was supposed to do and how to do it. Jeffrey shared the following:

Well, a lot of that knowledge came from the company that I am now part owner of. I might add that prior to being part owner of the company, the company's primary focus was project management consulting. They have since expanded into other areas. But a big part of the training I received was on the project management processes: the roles and responsibilities of the project team, the roles of a stakeholder, the role of a project champion, and the role of a project manager. That training gave me my widespread knowledge. However, the detailed knowledge and understanding has come through a refinement of that [widespread knowledge] through better understanding of what people's perceptions and expectations are.

I asked Jeffrey to elaborate on the training he received, and he said he received formal training through an organization that was a registered educational provider (REP). An REP is a company that is approved by the Project Management Institute (PMI) to provide training that project management professionals (PMPs) can use to accumulate and meet the

PDU requirements set by PMI. All PMPs must acquire 60 PDUs every three years to maintain their PMP certification.

Next, I asked Jeffrey how he learned to solve the problems that a project manager has to face on a regular basis. Jeffrey suggested that his problem solving skills were cultivated at a very early age. He shared the following:

> I would say that learning how to solve project management problems actually evolved through my childhood, and high school years. I learned primarily through:
>
> 1. Trial and error
> 2. Failed communication, failed risk management, or failed contingency plans.

I asked Jeffrey to describe the process of his growth from novice project management to his present status. Jeffrey shared the following:

> In my opinion that growth occurred through the understanding of the following:
>
> 1. What makes a project manager a good project manager and what defines success for each project. In my opinion, success for each project is defined differently depending on what the scope of the charter and what the actual objective of the project is.
> 2. An appreciation for the value of project management. I realized as an engineer that an organized project helped me do my work better, faster and more efficient.

3. An understanding of the schedule, how the schedule works, the dynamics of the schedule and how to look at the critical path and make a decision.
4. A process by which I could apply my project management knowledge in multiple industries and complete multiple projects on time, under budget and within quality standards.

Finally, when I asked Jeffrey to discuss the most important thing he has learned through his work as a project manager, Jeffrey cited perceptions. In many situations, perception can become people's reality. Jeffrey continued with the following:

I've learned to understand the importance of perception. A lot of people don't take the time to understand what others' perceptions are. They just assume since it wasn't their intention to communicate in a certain manner or it wasn't their intention to deliver information or execute in a particular manner, that it's okay. What they fail to understand is that other people's perceptions matter. So, in my opinion, the most important thing that I have learned to be aware of and understand is perceptions. Being cognizant of perceptions enable me to communicate with my team better, while coming up with the proper project execution strategy.

Continuing Professional Education (CPE): Its Role in a PM's Evolution

CPE is not a new concept. According to Queeney (2000), CPE was given its name and recognized as a component of adult education in the 1960s. Houle (1980) later wrote that expanding technology, rapidly growing knowledge bases, changes within professions, and the emergence of new professions clarified the need for more and more structured education for professional practitioners throughout their careers. Still, there are those like Azzaretto (1990) who argue that there is no guarantee that CPE participation will lead to better learning or improved practice. However, Jeffrey's comments may dispute that just a little bit.

I asked Jeffrey to share a set of events or activities that had the greatest impact on his professional development as a project manager. Jeffrey cited two situations. The first situation is discussed here, and the second situation is discussed in the section on informal learning. The event discussed here involves Jeffrey's first encounter with MS Project and the significance of its use.

Specifically, Jeffrey learned why he should use MS Project instead of MS Excel to create a project schedule.

> Early in my career as an engineer, I thought project management was more of a burden than a value-added methodology, and I just didn't understand why using Microsoft Project was important when you could quickly make a list of project activities in MS Excel, put the end date down and execute. But I later discovered the importance of using MS Project to create my schedule, when I determined that I could essentially use MS Project to create dynamic schedules that would allow me to predict whether or not a project would finish on time.

MS Project is a scheduling tool used in project management by project managers. Typically, MS Project is used for smaller projects as a scheduling tool, while Primavera is used for larger projects. The line of demarcation that dictates when to use one over the other is usually determined by the project manager or the consulting firm. Using MS Excel to create schedules nowadays is considered primitive at best.

Creating a work environment that is conducive to delivering projects successfully can be challenging, especially given the complexity of projects and the meticulous planning, preparation, and execution that is required to deliver projects on time, under budget and within quality standards. Therefore, I was intrigued by Jeffrey's feedback on the morale or atmosphere in his company, especially because he is president and part owner. I asked Jeffrey about the morale level and atmosphere in his company, and Jeffrey shared the following:

> I would say the atmosphere is very good ... I would describe it by saying the atmosphere is very energetic. We have a great group of people that really care about the deployment of our methodology as well as the execution of our client's projects. We've got a lot of very smart individuals ranging in experience from two to three years out of college to four years of experience, and to be able to see individuals work together regardless of experience level, it's just a very harmonious and energetic environment.

Informal Learning: Its Role in a PM's Evolution

Specific literature on informal learning emerged in the 1980s, but there is evidence of earlier research that can be attributed to Lindeman (1926), Dewey (1938), and Knowles (1970) who suggested that adult learners used self-direction as a gateway to their learning experiences. Table 2 and the literature showed that informal learning is synonymous with self-directed learning, coaching, and mentoring among other things. It also showed that informal learning does not usually take place in a classroom setting, but it can take place in the normal course of daily events with a high degree of structure.

Eraut (2004) expanded on Lindeman, Dewey, and Knowles by saying informal learning provides a straightforward contrast to formal learning and suggests a greater flexibility for adult learners. Eraut goes on to say that informal learning is implicit, unintended, opportunistic, and unstructured. Eraut's commentary was a good entry to my next interview segment with Jeffrey.

In the CPE section of this study, I asked Jeffrey what events or activities had the greatest impact on his professional development as a PM. I said one event would be discussed under CPE, and the other event would be discussed here under informal learning. Jeffrey shared the following informal learning event; it was basically borne out of self-directed learning:

> I learned through self-directed study how strong facilitation skills could benefit a project manager while he/she was completing a project. I had decided to learn more about facilitation skills, so I read through a couple of self-paced study guides, and documents what I thought would enhance my facilitation skills.
>
> Shortly after completing the self-paced study, I started applying some of the principles and techniques and subsequently realized how much more effective my meetings were and how much "waste" time was cut out of those meetings because of the information gained.

Incidental Learning – A Component of Informal Learning: Its Role in a PM's Evolution

Incidental learning is a category included in informal learning, and it is usually achieved through: Involvement, assumptions, beliefs, values, hidden agendas, the actions of others, and by learning from your mistakes. These are learning scenarios that typically take place outside of a classroom or an instructor-led environment.

Mumford (1995), as part of his work-based learning study identified incidental learning as one of four broad approaches to learning that are used either tacitly or deliberately. He said that incidental learning involves learning by chance from activities that jolt a person into conducting a post-mortem (investigation after the project is complete).

Jeffrey shared an example that qualified for this type of learning scenario. This incidental learning event was somewhat self-inflicted, but he learned a valuable lesson from it. The event involved an unsuccessful project caused by Jeffrey's failure to define and explain to his project team when an activity was considered complete. In essence, he did not explain before the project began when work on an activity should cease. As a result, the activity was not completed. Jeffrey summarized the scenario by sharing the following: "Another event that had a significant impact on a project's outcome actually led to a project's failure." The key factors that led to the project's failure were:

1. I failed to thoroughly define when an activity was considered complete, and therefore, there was no baseline to compare work against, and consequently, it led to incorrect updates and status reports on the progress of the activity.
2. I complicated the situation by taking people's word that certain elements of the project were complete. In addition, there were no checks and balances in place to alert us to what was going on.
3. I didn't have any clearly defined milestones, so I didn't know that the activity was not complete because there was no definition. One engineer thought design complete meant, "Hey, I've got the drawings done." The next engineer thought,

"No, the drawing has to be detailed, checked, and released." And the third engineer basically thought that, "No, you have got to have a prototype built, that's when the design is finished."

4. [In summary], because there was no clear communication, we got to the end of the project and failed to deliver the project. The sponsor lamented, "I don't have any of the deliverables." As a result, the project ran another eight weeks, and our team ended up being eight weeks late and about $150,000 over budget.

I asked Jeffrey to discuss the person or persons who have had the greatest impact on his professional development as a project manager. Jeffrey cited his relationship with his father, focusing on the incidental learning activities that facilitated the process of learning from the actions of others. Specifically, Jeffrey said he learned work ethic and perseverance from his father.

Formal Learning: Its Role in a PM's Evolution

As shown in studies 1 and 2, formal learning is typically learning sanctioned by an institution such as a college or business. In addition, it is usually highly structured in nature and should lead to credits, certification, or a diploma. Finally, this type of learning includes an instructor, curriculum, and an evaluation process (Hrimech, 2005; Merriam et al., 2007).

I asked Jeffrey to discuss the persons or events that have had the greatest impact on his professional development as a project manager. Jeffrey cited two examples. The first example took place in a structured institutional environment, and the second example took place in a coaching or mentoring environment. Both are classified as formal learning events.

> Well, for the first example, I have to go back to a point in time in high school, before I was formally consider a project manager. I had a teacher in high school that had a positive impact on my life … Back then, I wasn't the most studious person in the school, but it was primarily because things came easy to me. So, this teacher recognized that, and even though I was not in the top ten students in terms of grades, he put me in honors classes. The move worked, and I ended up excelling because the classes interested me, and they were very fast paced. Fortunately for me, he realized that challenging me and keeping me busy was the best way to motivate me and get the best out of me.

Jeffrey's second example involved a formal learning event in which he was mentored and coached by a business person who happened to be the former president of the company he currently works for. Jeffrey shared the following:

> The second example involves my current business partner. She has been an expert in project management for over 30 years, and she was actually involved with a lot of the people who founded PMI. At the time PMI was founded, she was working in Detroit and ended up collaborating with many of them during the early days to help them create PMI's project management methodology. As a result of her involvement with PMI, she has been the primary influence on the methodology that we currently use in our company today. More importantly, she has been my personal mentor, coach, and current business partner.

Open Theme – Jeffrey's Advice for Those Getting into Project Management

I asked Jeffrey to take a moment, review his career, and then answer the following question: What advice would he give to people trying to get into the field of project management? I added that this advice would be for a person who is not currently in the field of project management. Jeffrey suggested the following:

1. The first piece of advice that I would give them is to really focus on your resume. Everybody does project management to some degree. Everybody plans. I mean, inevitably, everybody has had a birthday party. Everybody has had an event. Everybody has had a gathering. Everybody has done something which required planning and organization. Look at that experience and document it. Don't overlook the things that you already have and the skill sets and experience that you have. Don't be afraid to use that as an example for possible PM employers.
2. Consider and really understand what it is that they want to do as a PM and how your current credentials stack up. But, if those credentials don't stack up, that's okay, don't give up!
3. Start learning, Start training. Take that opportunity to utilize an internship or something to gain that credibility in the industry that says, "I've got the experience. I've got the knowledge. I've got the ambition and I've got the understanding."
4. Consider charity work, the Boys and Girls Clubs, the United Way, and St. Jude. Experience and on-the-job training doesn't necessarily have to come from a paid position.

Jeffrey went on to say that those who are already in the field of project management but find themselves at the novice end of the project management novice-to-expert continuum should consider the following:

> Be aware of perceptions! Sometimes people will wrongfully assume a new/young person does not know what they are talking about. And they may say, "How could you possibly be a project manager of this project if you're only 28 years old?" Be aware of those perceptions and make sure that once you're aware of those perceptions, that you are able to address them and change those perceptions.

I closed the interview with Jeffrey by asking him if there was anything else he wanted to share with me about his development as a project manager. Jeffrey provided the following:

1. Absolutely! Always be willing to step outside your comfort zone. Just because you run a successful research and development project, it doesn't mean that you're not capable of leading a successful pharmaceutical project or IT systems installation project. Don't be afraid to step out of what you know in terms of project industry and to other industries and apply the principles that have made you successful.
2. Just because you're technically strong and can solve problems doesn't necessarily mean you will be a great project manager. With that said, go into each situation with confidence and the intent to learn and expand your skill set.

Jeffrey went on to discuss what he does specifically to sharpen his skills and/or maintain his craft.

1. Personally, I try to look at ways that we can make things even more efficient than they are. I asked questions like: How can we handle bigger projects? How can we utilize software and tools to our advantage?
2. I try to learn new technology and explore the benefits and limitations associated with it.
3. I want to emphasize the importance of not being afraid of new situation, stepping outside of your comfort zone.

In summary, the interview and study with Jeffrey was similar but slightly different in some ways from Brad's and Karen's. Jeffrey has an undergrad in Mechanical engineering, did not complete an MBA, but he is a PMI certified PMP. He ascended to his level of expertise through limited mentorship, formal training, structured learning events, a keen understanding of the importance of customer perception and the acquisition of PMP certification. More importantly, because he is a PMI certified PMP, he is required to initiate, cultivate and maintain a CPE agenda.

Jeffrey is unique in that he has been promoted to president of his company, since the interview, and can discuss and share the experiences from his journey, in the company, with the people he hires and manages.

Cross-Study Theme Analysis

This qualitative study involved an analysis of project managers in the context of the engineering procurement and construction management (EPCM) industry. Specifically, the study utilized three studies to complete a qualitative analysis of Brad, Karen, and Jeffrey in an effort to investigate how project managers become experts.

The focus of the study centered on project managers who were at various levels of competency along the novice-to-expert continuum and sought to ascertain through interviews answers to the following research questions:

1. How do project managers (PMs) in the EPCM industry conceptualize and define the role of an expert project manager?
2. How do project managers describe their evolution from novice to expert?
3. Are there any other factors that contribute to project managers becoming experts? If so, how are these factors related to the three types of expertise: absolute, relative, and the theory of deliberate practice?

The research questions are discussed in Chapter 5. The balance of this chapter will be spent reporting on cross-study comparisons using the themes from each study as a guide. The four major themes uncovered in the interviews and discussed in the previous chapter are project management, continuing professional education (CPE), informal learning, its subcategory incidental learning, and formal learning. This comparative analysis closes by discussing the open theme that surfaced as a result of open-ended questions utilized in the interview process. The open theme covers the question, what advice would you give someone who is trying to began a career in project management.

Project Management

One of the items I sought to uncover during this study was how project managers get their start in the field of project management. Brad, Karen, and Jeffrey began their careers in a manufacturing environment, and their introductions to project management were similar in that each of them was thrust into a situation where a project manager was needed. Basically, there was a need or requirement for someone to step up and do project management work. But their introductions were different in that Brad was already doing process engineering work when project work was thrust upon him. Karen was recruited from college into project management, even though at the time she knew very little about it. Jeffrey was hired as an engineer out of college, had already done project work in college, and was doing both engineering and project management work within three months of being hired as an engineer. The common denominator that appears to define their introduction is a project management need; the need, by their companies, for someone to assume the position of project manager.

Brad learned what a project manager does and how to do it through trial and error. During our interview, Brad made a point of saying, "I never had anyone mentoring me as far as project management goes, nobody showed me the ropes or gave me pointers." Karen, on the other hand, learned what a project manager does through a formal and incidental learning process. She attended formal project management courses at work, worked through structured practical learning scenarios, and had a project manager who worked as a mentor in the early part of her career. Jeffrey learned what a project manager does through formal learning.

The company Jeffrey works for is a registered education provider (REP) approved by PMI (Project Management Institute) to provide project management training that could lead to PMP (project management professional) certification. After completing this training, Jeffrey passed the PMP exam.

When I asked Brad, Karen, and Jeffrey to explain how they learned to solve the problems that project managers face, Brad and Jeffrey replied with similar learning processes. Brad learned to solve the problems that project managers face by observing other project managers and enduring what he calls "learned common sense." Brad defined "learned common sense" as "trial and error." Jeffrey indicated a similar method of learning, in that he learned from trial and error as well. Karen's learning process was somewhat different in that

she learned to solve the problems a project manager faces through a combination of CPE, formal learning, and informal learning.

Brad, Karen, and Jeffrey had opposing views on their journey from novice to expert (current status). Brad characterized his journey as one of trial and error, hard knocks, and extended time. He spent 18 years as a process engineer doing project management work before he was formally given the title project manager. Karen, on the other hand, experienced a planned and structured developmental process in her career, and although she asked a lot of questions, she always felt she had a support system. In her words, "the company I work for prefers to grow their project managers from the ground up, while indoctrinating them to their system early on and throughout their careers."

Jeffrey said the process of his growth from novice to expert has hinged on and been consumed by his understanding of five key components: 1) the value of project management, 2) what makes a good project manager, 3) what his team needs to be successful, 4) when a project is considered complete, and 5) what project success looks like.

Finally, I asked Brad, Karen, and Jeffrey to discuss the most important lesson they have learned through their work as project managers. Brad said, "I have learned you can't wing it." Meaning, you can't take short cuts and ignore project management methodologies and tools. Karen said the most important lesson she has learned is how to be a project manager. Specifically, she pointed to the importance of building relationships, understanding the roles that people play on a project, and maintaining self-confidence. Jeffrey cited the word "perception" as a word that has played an important part in the lesson he has learned as a project manager. Jeffrey suggested that far too many project managers fail to understand that other people's perception matters. He went on to say that close attention to the importance of perception can lead to better project communication and better project execution plans.

Continuing Professional Education (CPE): Its Role in a PMs Evolution

Brad, Karen, and Jeffrey each had CPE events that shaped their careers in some form or fashion. As a backdrop for the CPE question, I asked Brad, Karen, and Jeffrey to explain the event that had the greatest impact on their professional development as a project manager. Each of their events emanated from continuing professional development events, showing the importance of CPE in the development of a project manager.

Brad said that going through the process of becoming a licensed professional engineer (PE) had a dramatic positive impact on his professional development as a project manager. According to Brad, the PE exam is a two-part exam. The first part of the exam involves basic engineering questions from the other engineering disciplines such as electrical, mechanical, civil, structural, etc. The second part of the exam was on his discipline, chemical engineering. Brad said that passing the exam gave him the knowledge he needed to understand what the engineers in the other engineering disciplines do. More importantly, as a project manager, while completing projects that involve these disciplines, it has been extremely beneficial for him as a project manager to know what they do, and how they do it, and it has definitely made his job easier.

Karen's CPE event involved some tragedy, but she was able to learn something from the situation. She had attended classes on project cost management internally at her company and subsequently began work on a project in which she was responsible for cost. Suddenly, without warning, her mentor took ill and passed away. Initially, Karen's confidence was shaken and she experienced emotional trauma and anxiety. Fortunately, she was able to recover emotionally and gain something positive from the cost management training and the situation in general. Karen said the event also helped her understand the importance of roles on a project and how relationships can and should play out on a project. The situation also helped her understand the importance of speaking up to get the guidance and support she needed when things were not going right on her project(s).

Jeffrey's CPE event involved MS Project. MS Project is a scheduling tool that, if used properly, can help project managers' plan, sequence, add resources, and perform what-if scenarios on project activities for the purpose of delivering projects on time and under budget. Jeffrey explained that for a long time he had used MS Excel to plan his project activities and could not see the worth in using MS Project. Jeffrey said that early in his career, he felt project management was a burden rather than a value-added methodology.

Fortunately, Jeffrey decided to learn more about MS Project, and he learned that through MS Project, he could create dynamic schedules that facilitate what-if scenarios, and accurate predictions about project completion. Obviously, the benefits to him, his team, and his clients were huge. More importantly, Jeffrey could see an immediate return on investment.

Informal Learning: Its Role in a PM's Evolution

Brad, Karen, and Jeffrey experienced informal learning events that were, for the most part, associated with self-direction, coaching, and performance planning. Brad participated in a joint venture project that required him to wear many hats: engineering, process design, construction manager, project management, engineering manager, and training coordinator among other things. In Brad's words, "The project made me a jack-of-all-trades." In addition, it was a multiyear project that required international travel. Brad also stated that he has worked on more than 100 projects, and none of them required him to do the amount and type of work required by that particular project. Needless to say, the project went a long way towards providing Brad with the confidence and experience he needed to become a successful project manager.

Karen experienced similar developmental situations when she worked on shutdowns at her company. According to Karen, plant shutdowns provided significant testing and learning scenarios for engineers, project engineers, and project managers. In Karen's words, "If you don't learn something during a shutdown, you will probably pay for it later." Karen had the distinction of working two shutdowns, they happen about every three to four years, one as a project engineer and one as a project manager. As a project engineer, she was mentored and experienced many learning events. She said she learned more during her first shutdown, which was 30 days long, than she learned the whole year prior to the shutdown. The second shutdown, approximately four years later, was a little different. Karen had the title of project manager and was expected to perform at a high level with very little or no support. With that said, she continued her practice of asking questions when she felt she needed help. Karen said both situations had a dramatic impact on her professional development.

Jeffrey's informal learning event took place during a self-directed learning activity. He had decided to learn more about facilitation skills and how they could impact his ability to complete a project more efficiently. He read through a couple of self-paced study guides on facilitation, and subsequently applied the principles and techniques he learned to his project meetings. He said the benefits of the training were improved efficiency and effectiveness in meetings.

Incidental Learning – Component of Informal Learning: Its Role in a PM's Evolution

Brad, Karen, and Jeffrey experienced incidental learning events that can be characterized by the process of learning from the actions of others, learning from involvement, and learning from mistakes. All three of these examples are listed as incidental learning in Table 2. Brad recalled a situation in which he was a plant engineer working with a project manager who Brad called a "force of nature." Brad said he learned a lot about efficiency and effectiveness while watching and working with this particular project manager. He said he was impressed by the project manager's uncanny ability to focus on only those things that were important, pushing aside those things that meant very little or had no value.

Karen shared an incidental learning event that involved one of her college professors in her undergraduate program. Karen said he was an experienced construction management professional, and he tried to teach the class how things were done in the real world. Unfortunately, Karen said she did not realize the importance of what he was saying until much later in life when she got a chance to actually see it for herself in practice. However, what he said came back to her as she was applying construction management principles during a work activity. In addition, Karen cited another event that takes place naturally with her coworkers as a result of the way their office space is configured. She said because they sit in close proximity to each other, communication and collaboration on problems is commonplace and typically results in the timely resolution of problems.

Jeffrey shared an incidental learning event that can be classified as an event in which he learned from his mistakes. While assigned as the project manager for a particular project, he failed to define when an activity was considered complete. Therefore, there was no baseline for the activity, and all progress reports on the activity were skewed or incorrect. In addition, Jeffrey complicated the situation by not checking the status of the activity, instead relying on the project team to report on project status. Ultimately, Jeffrey did not find his mistake until the project was coming to a close. The sponsor asked for a deliverable associated with the activity that had not been defined, and it was then that Jeffrey realized the error. This oversight caused the project to be eight weeks late and result in a budget overrun of $150, 000.

Formal Learning: Its Role in a PM's Evolution

Brad has had very little formal project management education over his career. His MBA was completed with a general management emphasis, and he does not feel it helped him with project management. Moreover, asked about his completion of formal project management education through coursework, seminars, or webinars, he replied, "I may have taken a one-day project management course 20 years ago." Brad went on to say he has recently felt the need to pursue PMI's PMP certification and as a result, took a PMP prep course to help him prepare for the exam.

Karen has been exposed to what I would describe as a structured and planned approach to her development as a project manager. She had formal training early in her career, and coursework coupled with specific project learning events over the years helped reinforce what she learned in the classroom. In addition, she completed an MBA at one of the top business schools in the U.S., but she does not feel the MBA has benefited her in project management. Karen considers what she learned in civil engineering as an undergraduate to be more beneficial to her project management career than what she learned in her MBA studies. With that said, Karen agrees that PMI's PMP certification is a worthwhile goal, and she plans to pursue certification to top off her career and find out what she does not know about project management.

Jeffrey also had the benefit of formal training early in his career. The company he worked for right out of college sent him to formal project management training as a precursor to sitting for the PMP exam. Although he does not have an MBA, Jeffrey has had to pursue project management training on a regular basis, for many years, to maintain his PMP certification. All PMPs are required to accumulate 60 PDUs every three years to maintain their certification. The project management institute (PMI) maintains a list of several categories on its website, pmi.org, that explains how a PMP can go about acquiring the 60 PDUs that are required to maintain their certification. Failure to acquire the 60 PDUs in the three-year timeframe could result in a PMP forfeiting his/her certification, in which case they would have to take the PMP exam over.

Open Theme: Advice for Those That Desire to Get into Project Management

I asked Brad, Karen, and Jeffrey a two-part question that called upon their experience and expertise. The first part of the question required them to reflect on their careers and explain what advice they would give to someone trying to initiate a career in project management. The second part of the question required them to provide advice to a novice project manager who had launched a career in project management and was looking to enhance their skills, while progressing toward the expert level on the novice-to-expert continuum.

Brad suggested that a newcomer to project management should pursue formal project management education, find a mentor to help them through the rough spots early in their career, and spend some time understanding what their project team does in performing their jobs. For example, learn what a developer does, learn what an engineer does, learn what a network administrator does, etc. Finally, initiate and cultivate a CPE agenda and make a conscious effort to practice the tools and skills you learn, until they become "learned commonsense."

Brad went on to say that those who are in project management, but find themselves at the novice end of the novice-to-expert continuum, should do pretty much the same as the person breaking into project management. More specifically, they should learn what project management is, spend time observing what other people do, read books, secure a mentor, practice your craft, and work hard.

Brad closed by explaining what he does to maintain and sharpens his craft. Basically, he said he spends time trying to perfect the use of the project management tools that are available to him and the profession. Brad had also initiated the PMP process and hoped to take the PMP exam in 2014.... Brad passed the PMP exam in 2015.

Karen suggested that a newcomer to project management should get some formal training and get a firm understanding of what project management is all about. In addition, she recommends that a new person pursue PMP certification because many companies are now requiring it. Finally, Karen suggested that a new person initiate and cultivate a CPE agenda.

For a person who is already a project manager, but finds themselves on the novice end of the novice-to-expert continuum, Karen recommends that they try to get as much project experience as possible while exploring the tools that are available to project managers to manage projects. Finally, Karen said it is important to build relationships. As a novice project manager, a person will need a network or support system that he/she can draw from.

Karen closed by saying she maintains and sharpens her craft by pursuing work on larger projects. She believes increased project complexity, coupled with an active CPE agenda, the use of project management tools, and a wellness program will go a long way toward helping her maintain and sharpen her craft.

Jeffrey said a newcomer to project management should gain as much practical experience as possible. He went on to say that a new person should seek project experience before launching a project management career. He said the experience can be gained through internships, and charity work with organizations such as the Boys and Girls Clubs, the United Way, and St. Jude. According to Jeffrey, there is no law that says you must get your project management experience from a paid position only.

Jeffrey went on to say that those who are in project management but find themselves at the novice end of the novice-to-expert continuum should be cognizant of the power of perception. According to Jeffrey, many times, a novice project manager will be ignored or cast aside because they are new to field. He recommends that novice project managers focus on those situations and aggressively seek to minimize, if not eradicate, the perceptions that surround those situations, because they can derail or stall a novice's growth and advancement along the novice-to-expert continuum.

Jeffrey closed by saying he maintains and sharpens his craft by looking for ways to be more efficient and effective in delivering projects. He also spends time trying to figure out how he can handle bigger projects and how he can integrate new software and tools that will make him a better project manager. As a final thought, Jeffrey suggested you can't be afraid to learn new technology and explore the benefits and limitations associated with it.

Assertions and Generalizations

Based on the information uncovered by this study, it appears that engineers and others get their start in project management because of a situational/company need. The nature of the work dictates that people who are familiar with a component of a project may, by default, wind up getting assign to complete the work. In the process, sometimes they are forced to learn on the fly, as Brad did. Other times, they may be fortunate enough to be provided with formal training and a mentor, like Karen, or just formal training, like Jeffrey.

CPE should be an important part of a project manager's life. However, Jeffrey, who is a PMP, is the only one in the study who has a clear and concise CPE agenda. This may be attributed to his obligations to PMI as a certified PMP. Brad (Brad passed the PMP exam after the interview in 2015), but even though Karen believes the PMP certification is important, she has not passed the PMP exam, nor has she committed to a CPE agenda; at the time of the interview. I believe pursuing and passing the PMP exam is a hurdle that must be cleared to be considered an expert project manager. Further, I believe companies in the EPCM industry should take the lead and adopt an aggressive PMP program which requires all project managers to pursue and achieve PMP certification.

Informal learning and its subcategory incidental learning also play an important role in the life of a project manager. Both learning scenarios, from my experience, take place on the practical side of project management. Things like self-directed learning, networking, coaching, learning from mistakes, learning from involvement, and learning from the actions of others are examples of pragmatic scenarios in which the project manager must extend him/herself beyond the classroom or institutional learning environment and make an effort to learn something that will benefit them in the field of project management. This pragmatic component cannot be taken for granted because it is an essential ingredient that, if ignored or not carried out properly, can preclude the growth of a project manager.

Formal learning has its place, but there is no proof that formal learning by itself translates into enhanced job performance. Therefore, formal learning must be coupled with the practical side of learning to approach meaningful learning. The approach that was used in Karen's development is, I believe, a good blueprint for learning.

Last, for those who are contemplating project management as a career, spend time understanding what project management is all about and take the time to acquire the proper skills, knowledge, and experience. Jeffrey did a nice job of listing the alternatives and available options to consider. Project management experience can be gained through pro bono experiences, and I recommend it for those that are trying to break into the field.

Project managers who are already in the field and want to improve their project management acumen should do the following: Complete the PMP certification process, initiate a CPE agenda, seek out a mentor, and seek out projects that will expose them to different aspects of project management. The more projects you are involved in, the better you become at your craft. After you have been a project manager a while and you begin to feel like you have reached saturation in terms of knowledge, spend some time doing research to compare methodologies, tools, software, and other items associated with project management. In addition, learn how project management is done in other industries. Currently project management is employed in the following:

Engineering, IT, Banking, Real Estate, Healthcare, Accounting, Sales, Finance, Advertising, Marketing, Telecommunications, Steel, Transportation, and Academia to name a few. Project management will typically be the same across industries, but the domain knowledge in the industries is something that can require a project manager to stretch themselves and become more adroit at. Summarily, there is always something new being developed that can impact the way you manage projects. As project management practitioners, we should all adopt a posture of lifelong learning.

Chapter 5

DISCUSSION

Summary of Findings

THIS STUDY WAS an investigation of the project management profession in the context of the engineering procurement and construction management (EPCM) industry. The conceptual framework of this study was based on four components: 1) expertise, 2) project management, 3) continuing professional education (CPE), and 4) informal learning. Although project managers exist and operate in many industries such as information technology, real estate, banking, pharmaceuticals, marketing, sales, advertising, and engineering, little is known about the steps, path, or journey of project managers from novice to expert.

Participants in this study consisted of three project managers (PMs) whose experience levels range from five to 30 years in the field of project management. The PMs were selected through convenient sampling from a pool of project managers that I had encountered over the past 25 years. The PMs interviewed for this study had the following criteria: at least one of the PMs had formal project management education and was recognized as a PMI-certified PMP; the second PM had formal project management education, but was not a PMI-certified PMP; and the third PM had no formal education and was not PMP certified. In addition, at least one of the three persons interviewed was a female project management professional.

A qualitative study was used to investigate specific information about the experiences, values, opinions, and social context of each project manager within each study, while a comparative approach across the three studies was employed to enable a thorough investigation and exploration of the phenomena that existed across the three studies in the context of the EPCM industry. All of the data were collected through the interview process. A list of eight interview questions can be found in Appendix A of this study. Each participant was interviewed for approximately 45-60 minutes, and all interviews were transcribed in Microsoft Word by a professional transcriber. After transcription, interview transcripts were mailed back to each participant to ensure that the data derived from the transcripts were congruent with the participant's original thoughts.

Comparison of the data derived from the interviews led to the identification of themes that helped to characterize the interviews. The connections between themes were derived by analyzing the transcripts from the interviews and my research questions. The themes uncovered were: project management (PM), continuing professional education (CPE), informal learning, formal learning, and a category that I labeled open themes.

I asked another person to perform a peer review of my research for inter-rater reliability. The participant interviews provided rich and varied descriptions of how PMs in the EPCM industry conceptualized and defined the role of a project manager, how project managers described their evolution from novice to expert, and the factors that contribute to project managers becoming experts.

The next section provides a connection between the research questions and the findings of this study and is organized by research question. I will list the research question and then discuss my findings from the study, explain what those findings mean, compare and contrast the information gathered from participants – Brad, Karen, and Jeffrey – with the literature reviewed in Chapter 2, and then explain whether I agree or disagree.

Relationship between the Research Questions and the Findings

The Role of a Project Manager

Question #1: How do PMs in the EPCM industry conceptualize and define the role of an expert project manager?

According to the literature, Pandya's (2014) position on the role of a project manager can be summarized by saying project managers are liable for the day-to-day supervision of the project, specifically the triple constraints of scope, time, and cost. However, they are also responsible for managing change, resource readiness, conflict resolutions, and emotional flare-ups with internal and external stakeholders, and relationship building that could aid in the creation of a high performance team (Pandya, 2014).

Dubois, Koch, Hanlon, Nyatuga, and Kerr (2015) stated that project management is an evolving practice and the project manager, leading the projects, has an important role of overseeing the project and project team, and ultimately ensuring the project ends in success.

The participants in this study – Brad, Karen, and Jeffrey – stated that there are several characteristics that make a project manager successful, and the comments by the participants are consistent, in some areas with the literature. First of all, a project manager must be able to galvanize and lead an unfamiliar, diverse group of highly educated professionals for the purpose of solving a unique, complex problem in a dynamic setting while delivering a project on time, under budget, and within quality standards where change is inevitable and must be managed and controlled. Second, a project manager should initiate and perpetuate a continuing professional education agenda. Third, a project manager must engage in project management activities that will lead to professional development.

Brad conceptualized and defined the role of an expert project manager by using the words leadership, delegation skills, knowledge, people skills, and ethics. He put a lot of importance on a project manager being adroit at leading a team of professionals, being able to delegate work when necessary, and being extremely knowledgeable. He also stated that having people skills and being able to get along with your team as an important aspect as well. He closed by saying that ethics in a leadership role is vital and that project managers should lead by example.

Karen conceptualized and defined the role of an expert project manager by highlighting the complexity of a project and the dollar amount associated with it. According to Karen, the dollar amount of the project sometimes determines the complexity of the project. In addition, span of control or the number of people on the project team is also an important component of what separates an expert project manager from a novice.

Karen went on to explain that senior project managers, like herself, should know the tools used by project managers, but, in her opinion, they become secondary and are less important than aptitudes like communication skills, leadership, and delegation skills.

Jeffrey had a similar opinion on how expert project managers are conceptualized and defined. Jeffrey defined an expert project manager by the number of successful projects he/she had completed and not by the number of years he/she had been a project manager. Therefore, from Jeffrey's point of view, he could see himself hiring a person with four years of project management experience instead of a person with 15 years of project management experience, if the person with four years of project management experience had successfully completed more projects than the person with 15 years of project management experience.

Jeffrey went one step further. He believes an expert project manager is defined by his or her ability to manage projects in different industries. For example, if your primary expertise is banking, Jeffrey believes you become or reach the status of expert when you can successfully manage projects in another industry such as real estate or engineering.

It has been my experience that the role of a project manager sometimes changes based on the needs of an organization. There have been times, while working as a project manager that I was asked to perform the work of one of my team members. Other times I found myself in the managerial role that Pandya (2014) described above. In summary, I agree with Pandya that a project manager will carry out numerous activities while working in the role of a project manager.

In addition, based on the literature and the findings of this study, expertise, as it relates to project managers and their role in 2017-18, is a combination of serial problem solving and precise professional development activities. Serial problem solving, according to Newell and Simon (1972), characterized the method of problem solving in the first generation, while precise professional development activities characterized the second-generation studies. Serial problem solving can be defined as a problem-solving technique or process that requires the investigation and review of a series of related, successive events that could lead to a problem's solution. Typically, serial problem-solving project managers will be called upon to solve problems, and they should have been exposed to precise professional development activities that prepare them for the plethora of project related issues they will encounter. The precise professional development activities were discussed in my findings from Chapter 4. Some of those activities include informal learning and its subcategory incidental learning. Both learning scenarios, take place on the practical side of project management. Events like self-directed learning, networking, coaching, learning from mistakes, learning from involvement, and learning from the actions of others are examples of pragmatic scenarios in which the project manager must extend him/herself beyond the classroom or institutional learning environment and engage in initiatives that will benefit them in the field of project management.

According to Daley (1999), those who used serial problem solving in the first generation and supplement it with precise professional-development activities demonstrated professional growth in their chosen career as they gained experience within the context of their work setting.

The literature provides a definition of expertise, and this definition will help compare this study's findings with the literature to see how this study's participants' views of expert project managers differ from the literature. According to the literature, the definition of expertise and how it is developed can be broken into three main theories: absolute expertise, relative expertise, and the theory of deliberate practice.

Simonton (1977) says absolute expertise is a result of chance and unique innate talent. The nature of project management will not allow, the use of absolute expertise as a measuring stick for project management expertise. Chi (2006) defined relative expertise as something that can be achieved. In fact, Chi believed that a less skilled person could achieve expertise and that relative expertise was an approach that helped facilitate that process. Ericsson (2006) defined the theory of deliberate practice as a process by which specific aspects of a person's performance, in this case project managers, are improved, measured, and integrated into their careers so that they can become better in their careers; meaning, better project managers.

After reviewing the literature, understanding the different types of expertise, and contemplating the comments from participant, I conclude that project management expertise in the EPCM industry can be conceptualized and defined by a combination of two of the three theories of expertise listed above: project management and what project managers do can be characterized as a relative expertise gained through deliberate practice. I made this assumption early in the literature review before interviewing the participants in the study. Unpacking this assumption led me to the conclusion that relative expertise makes the difference between novice and experts, and that is the case in project management.

Deliberate practice tries to measure performance and improve on it through a series of activities on the way to mastery of weaknesses and perfection of acquired skills (Ericsson, 2006). Many of the comments made by participants focused on getting more experience through leading more complex projects with larger budgets and wider spans of control.

However, it has been my experience that there are additional activities that lead to improved performance. For example, activities that qualify as performance-enhancing activities in deliberate practice would include project activities that present challenges in the areas of diversity, culture, geographic displacement, new product development, and leading edge technological advancements. Each of these items requires difference levels of experience and expertise, the scope of which is beyond the boundaries of this study.

According to Walker (1997), organizations hire project managers, full time or part time, for reasons that include: 1) to acquire expertise that has otherwise been unavailable, 2) to secure unbiased perspectives and professional recommendations, 3) to maximize cost-efficiency and cost effectiveness, and 4) to supplement their existing management team for a short period of time.

In addition, with any project, it is important that project leadership be assigned to project managers familiar with the manufacturing processes and culture of the company in question. Adhering to this process and applying such expertise improves the effectiveness and efficiency of the project through unbiased support and the skilled application of project management tools and techniques (Brown, 2000).

The participants in this study had similar comments about the role of a project manager and the importance of leadership skills, experience gained through the management of multiple projects, mastery of project management tools and techniques, the ability to communicate, and familiarity with company culture.

Summary

The role of a project manager is characterized by the effective and efficient supervision of four major components: scope, time, cost, and quality. However, a project manager is also responsible for overall project success, but the path to overall project success can take a number of different turns, including project overruns in time and cost, deviation from customer quality requirements, delays in project equipment and/or material, unforeseen problems, or problems related to risk. In addition, the path to project success will require the project manager to demonstrate a number of different skills, for example, leadership skills, delegation skills, communication skills, technical skills, interpersonal skills, emotional intelligence, mastery of project management tools and techniques, and initiation and perpetuation of a CPE agenda that will require the project manager to constantly engage in professional development activities.

Novice to Expert: A PM's Journey

Question #2: How do project managers describe their evolution from novice to expert?

The journey from novice to expert is one the literature classifies as the theory of relative expertise. Chi (2006) explains it this way: expertise is a level of proficiency that novice can achieve. However, an organized body of domain knowledge is a prerequisite to expertise and, by definition, experts possess a greater quantity of domain knowledge than novice.

It is important to note that there is an organized body of knowledge for project managers to use as a reference, and it is called the project management body of knowledge or PMBOK.

Brad gained his level of expertise through the school of hard knocks. He had very little formal education and acquired the majority of his knowledge, skills, and training through informal learning and its subset, incidental learning. His journey can be explained by the following words: lengthy, time consuming, deliberate, and learned commonsense.

Karen's journey from novice to expert can be characterized by baby steps in formal education, on-the-job training, and working in different roles on diverse projects. Karen's journey was different from Brad's in that Karen had formal education, structured on-the-job training, a mentor, and a support system in the form of her coworkers. She felt empowered by the ability to ask questions, and she also took the time to build relationships, which is always a good thing. Karen's journey was similar to Brad's in that it took time to become the project manager that she is, and, although she did not use the words school of hard knocks or learned commonsense, she experienced hard knocks and trial and error during her journey. She cited one situation in which her mentor died at a crucial point in her development.

Jeffrey's journey from novice to expert can be described as one highlighted with pragmatic or hands-on experiences. In addition, he had formal education and a mentor like Karen. Like Brad and Karen, his progression along the novice-to-expert continuum took time, but unlike Brad and Karen, Jeffrey had the benefit of being a PMI-certified PMP.

As a PMI-certified PMP, Jeffrey had a framework and body of knowledge, via the PMBOK, that provided and continues to provide a structured approach to how certain project-related activities can be handled. Without this framework, Brad experienced the school of hard knocks and Karen experienced baby steps. I believe the PMP certification gave Jeffrey an advantage over Brad and Karen in his journey.

During his journey, Jeffrey viewed each project as a piece to a puzzle or a bigger picture. Stated another way, he described his journey as one in which he viewed each project as a stepping stone to greatness or expertise. But the expertise he was referring to includes a journey outside of the EPCM industry into other industries. He believed that a project manager should be considered an expert only when he/she has managed projects outside of their current industry. For example, if you are currently working as a project manager in the banking industry, you can only achieve the level of expert project manager if you successfully manage projects in real estate, healthcare, information technology, or engineering. To a certain extent, I agree with this statement. However, I would caution Jeffrey and anyone else to be careful when they say you are not an expert project manager unless you have managed projects in more than one industry. I believe it is reckless and somewhat irresponsible to take that position. As a member of the Project Management Institute (PMI), I have met and networked with people who have worked in one industry all of their lives, and I believe they are good at what they do.

For example, the executive vice president of a local engineering procurement and construction management (EPCM) company (Company B), has spent the majority of his 20-plus-year career at one company, and he indicated (in a personal communication, 2014) that project managers in his company were considered competent and/or expert project managers if they could push a project(s) through the gauntlet that exists inside the company, and deliver those projects to the customer (outside) of the company on-time, under budget and within scope requirements.

By comparison, Grochow (2008) said it takes two to three years of experience for a person in the nursing profession to reach the competent stage. From that point, ongoing education and support are needed for continued progression to the expert level. Gruchow did not equate competency to a specific position on the novice-to-expert continuum, but she did indicate that competence was a prerequisite for achieving progression along the novice-to-expert continuum.

White (2015) stated that the progression from novice to expert in the nursing profession can be characterized by a three step process: 1) introduction of an assigned mentor, 2) supervision of task given to a novice nurse and feedback and support as the nurse progresses along the novice-to-expert continuum, and 3) the "letting go" process. Although, according to White (2015), the mentor never really lets go, even after a nurse achieves the status of expert.

Grochow (2008) and White (2015) describe activities that are consistent with the methods used in the profession of project management. Specifically, it takes time to progress from novice to expert in project management, as evident by Karen's "baby steps," Brad's "school of hard knocks," and Jeffrey's "piece to the puzzle" experiences.

This research question placed participants in the context of relative expertise. They were asked to describe a journey that has the goal of relative expertise, which is to help less- skilled or experienced people achieve the title or position of expert (Chi, 2006). Each of the participants described, at varying degrees, the steps they took in the theory of deliberate practice to achieve their level of expertise. The steps taken by the participants in this study support the theory of deliberate practice in our conceptual framework. However, the specifics of how project managers reach the level of expertise are indigenous to the project management profession. Some of the steps in project management may apply across other professions, but some may not. *The literature on relative expertise helps us understand how to explain the difference between novice and expert. It also provides a basis for us to begin the dialogue or discussion on the steps required to achieve expertise through the theory of deliberate practice.*

Summary

The journey of a project manager from novice to expert can be an arduous one. The participants in this study used a variety of terms to explain their journey, and it is important to note that each person (participant or otherwise) can take different paths to the same destination; in this case, the destination is expertise.

Brad used words like lengthy, time consuming, deliberate, and learned commonsense to describe his journey. Brad had very little formal education and had not achieved PMP certification when interviewed.

Karen characterized her journey by using the words "baby steps." She said she took baby steps while engaging in formal education, on-the-job-training, and working in different roles on diverse projects.

Karen's journey was different from Brad's in that Karen had formal education, structured on-the-job-training, a mentor, and a support system in the form of her coworkers. She felt empowered by the ability to ask questions, and she also took the time to build relationships, which is always a good thing.

Jeffrey's journey from novice to expert can be described as one laced with pragmatic or hands-on experiences. In addition, he had formal education and a mentor like Karen. Like Brad and Karen, his progression along the novice-to-expert continuum took time, but unlike Brad and Karen, Jeffrey had the benefit of being a PMI-certified PMP. As a PMI-certified PMP, Jeffrey had a framework and body of knowledge, via the PMBOK, that provided and continues to provide a structured approach to how certain project related activities can be handled.

Maintaining Expertise

Question #3: Are there any other factors that contribute to project managers becoming experts? If so, how are these factors related to the three types of expertise: Absolute, relative, and the theory of deliberate practice.

The theory of deliberate practice is considered a development domain because the goal of deliberate practice is to improve specific aspects of performance in such a way that attained levels of proficiency can be measured and integrated into a person's future performance (Ericsson, 2006). A person in the profession of project management, and any profession for that matter, should seek to improve him/herself. Improvement can come in many forms. Project managers who achieve PMI's PMP certification are required to secure 60 professional development units (PDUs) every three years to maintain their certification.

Brad acknowledged that formal education is a factor that could contribute to the journey of project managers from novice to expert. He also suggested that successful execution of diverse projects with gradual increases in project complexity is a contributing factor. As far as the theory of deliberate practice, Brad believes his pursuit of PMI's PMP certification falls in this developmental domain. He said he likes PMI's requirement of 60 PDUs every three years as a means of keeping abreast of the latest tools, techniques, and methodologies in project management. He said it forces a PMP to keep his/her knowledge and skills current.

Karen responded to this question by saying cultivation of leadership skills can contribute to a project manager becoming an expert, and then she tied leadership to the theory of deliberate practice. In Karen's opinion, leadership must be practiced, measured and practiced again to realize measurable improvements. She also cited mentorship, a component of deliberate practice, as a factor that could contribute to a project manager becoming an expert.

Karen also acknowledged that the pursuit and maintenance of PMP certification would go a long way toward solidifying and enhancing a person's career and status as a project management expert. She believes the PMP activity would fall in the theory of deliberate practice domain.

Jeffrey, who is a PMP, suggested that continuing professional education (CPE), formal, and informal learning could contribute to a person's journey toward expertise. Because he is a PMP, he subscribes to PMI's 60 PDU requirement as a great tool for lifelong learning and professional development. Jeffrey also cited a factor that Brad utilized a big portion of his career, and that factor is, learning through observation. Observation, according to Brad, that is planned and structured can contribute to a project manager's expertise. In addition, Jeffrey said a person's ability to communicate is a key component, and a contributing factor in a project manager's quest to become an expert.

Jeffrey concluded this portion of the interview by linking the concepts: communication and observations. He said a person must practice the acts of communicating and observing. He then suggested that communicating could be an absolute expertise for some people. Not all contributing factors, in Jeffrey's opinion, are components of the deliberate practice domain. According to the PMI (2005), a project manager (PM) will spend 90% of his/her time communicating. Therefore, it is important that a PM be adroit at communication, both written and oral. Whether communication is an absolute expertise, as Jeffrey alludes to, may or may not be true, but it is beyond the scope of this study.

In addition, a PM must be a meticulous observer because one of the ways a project manager can evolve from novice to expert is through informal learning and its subcategory, incidental learning. Incidental learning involves learning from mistakes, learning from involvement, and learning from the actions of others, i.e., observing.

I believe a person should, at some point in their project management career, sit for and pass the PMP exam. After passing the PMP exam, PMI requires you to perpetuate a CPE agenda by requiring 60 PDUs every three years. These requirements, in my opinion, open the door to a required lifelong learning agenda that I believe is central to the deliberate practice theory of expertise. The core assumption of the theory of deliberate practice is that expert performance requires the meticulous search for training and tasks that a person can master sequentially.

I will close by saying, project managers who are cognizant of the benefits of PMP certification and take advantage of those benefits after becoming a PMP will find themselves on the road to expert project manager and the learning path that the theory of deliberate practice purports as its goal in the expertise domain: which is, to improve specific aspects of performance in such a way that attained levels of proficiency can be measured and integrated into a person's future performance (Ericsson, 2006).

Summary

After achieving the level of expert, a project manager must seek to maintain that level. The participants in this study discussed a variety of ways to accomplish that feat. Brad cited formal education and completion of the process that leads to PMP certification. Karen suggested that a continued focus on the improvement of leadership skills, the support of a mentor, and completion of the process that leads to PMP certification as activities that contribute to the maintenance of expertise. Jeffrey, who is PMP certified, advocates formal and informal education along with the proper use of observations as a means for maintaining expertise. Jeffrey also suggested that continued improvement in the area of communication, both written and oral, are essential to a project manager's continued success.

Discussion

During my (several-year) tenure as a consultant/ project manager with three engineering firm in the western suburbs of Chicago, I contemplated my dissertation topic and the salient issues that plague the profession of project management. I saw many inexperienced, and some experienced, project managers struggle to deliver projects that were consistent with clients' request. Therefore, I decided to interview company executives who could give me their candid assessment of the ubiquitous issues that plaque them and the companies they manage.

I conducted four semi-structured interviews with executive-level professionals of well-known EPCM companies in the Chicagoland area: I will refer to them as Company A, (two executives from Company A), Company B and Company C. As I mentioned earlier, I was employed at Company A.

The first executive I interviewed at company A (personal communication, December 3, 2013-14) stated that he had recently read a scope statement written by one of Company A's senior project managers, and he was appalled at the contents of the document. The first executive cited sentence structure and grammar as major issues. In addition, Company A's first executive said project managers fail to understand the importance of the triple constraints: scope, time, and cost. Managing these three items are key components of project success. Finally, Company A's first executive said project managers must possess industry knowledge and a variety of skills to be successful, including the ability to communicate with irate clients and the ability to manage change.

Company A's second executive (personal communication, December 3, 2013-14) cited many of the issues highlighted by Company A's first executive, but also cited conflict resolution as an important issue. Company A's second executive went on to say some project managers lack the personality to be effective project managers and closed by saying some project managers didn't fully understand the work flow at their company.

Company B's executive (personal communication, December 3, 2013-14) suggested that project managers at his company have three issues that affect their performance. Those issues include project managers' inability to write an effective scope statement, their lack of project management knowledge and their inability to write a definitive, holistic, action-oriented project execution plan (PEP). Company B's executive also cited their inability or unwillingness to manage the project team, along with their lack of teambuilding skills. Finally, he said project managers should not perform technical work. Technical work should be delegated to the appropriate team member. For example, electrical engineering work should be completed by electrical engineers, not project managers.

The executive at Company C (personal communication, December 5, 2013-14) suggested time management is a major issue for project managers. Company C's executive continued by saying project managers should be adroit at reading schedules, whether it is in Primavera or MS Project format. He also cited planning and the lack of diverse knowledge as an issue. According to Company C's executive, a project manager must be a well-rounded, knowledgeable professional who can integrate all components of a project. An understanding of "earned value and / or earned value management" is also an important concern.

I believe the majority of the above issues, can be addressed by CPE, or project management professional (PMP) certification. Project Management is a relative expertise that is gained through deliberate Practice." Relative expertise characterized the type of expertise I saw, experienced, and worked through in my seven-year tenure as a consultant/project manager. Many times an experienced project manager – with experience typically measured by the number of projects and/or number of years with the company – was paired with an inexperienced or novice project manager. The latter of the two was expected to learn from the former. Unfortunately, that is a slow and inefficient process for the type of work required in the industry. More importantly, Rockhill (1983), Darkenwald and Merriam (1982), and Lisman (1980) recommend mandatory CPE, citing its importance to each profession and a step in the right direction, if not the cure for maintaining competency and proficiency. In addition, Ericsson (2006) suggested that there are several factors that influence the level of professional achievement or expert status. Those factors include experience, training, formal education, supervision by more experienced professionals, and regular, prolonged execution of activities that characterize the professional domain that you work in (p. 683); in this case, the project management domain.

Therefore, no longer should the EPCM industry continue with business as usual, exhibiting a passive approach to CPE, formal training, and the PMP process. History has shown, and the EPCM industry has agreed, continuing in the same direction will result in further sub-par project performance.

This study uncovered pertinent information about the three paths project managers can pursue to become experts (shown below); *I believe the EPCM industry should consider each path and aggressively pursue path number three (below):*

1. Achieve expert status through on the job training or experience, informal learning, learned commonsense, and the school of hard knocks; this is a viable but long and arduous path, as pointed out by Brad in this study.
2. Achieve expert status through formal education, structured learning events, formal training, and mentoring and/or coaching; this is a quicker path than path #1 and it provides more support and less frustration.
3. Achieve expert status through structured learning events, formal training, mentoring and / or coaching, the acquisition of PMP certification and a CPE agenda; I believe this is the most appealing and efficient path.

There may be other ways that I did not uncover or discuss but the participants – Brad, Karen, and Jeffrey – in this study provided much of the insight captured in this discussion, some of which I did not necessarily believe or agree with when I began my research, but I absolutely agree with now.

I want to further point out that it is very important that the EPCM industry aggressively pursue a PMP and CPE agenda that will lead to their project managers being better educated and prepared to lead and manage projects. PMP certification will provide project managers with a foundation that is essential to continued and improved delivery of successful projects.

The reasons being, before you can sit for the PMP exam, you must meet two criteria: 1) you must show that you have successfully managed projects for at-least 2-1/2 years, and 2) you must show that you have formal training in project management. Finally, you must pass a 4-hour 200-question exam. These three components, successfully completed by a person, says a lot about a person's foundation in project management and their readiness to lead and manage projects. As I discussed in the study, Brad's path was characterized by hard knocks, and Karen experienced similar hard knocks, but to a lesser degree. However, Jeffrey, who pursued and achieved PMP certification early in his career, did not experience the hard knocks or other negative consequences that Brad and Karen did. More importantly, I believe completing the PMP requirements is a necessary step, along with CPE, toward the coveted status (title) of expert project manager.

Contributions to the Literature

Little has been written about the journey or outcomes of project managers in the EPCM industry as they travel the intellectual path that represents the novice-to-expert continuum. Moreover, few, if any, studies have linked the experiences of project managers (PMs) in the EPCM industry to the literature constructs of expertise, project management, CPE, and formal and informal learning. Although this study is not all encompassing or valid for all project managers in all industries, it provides important insight into the experiences of three project managers via studies in the context of the EPCM industry.

Understanding and appreciating the studies of the participants, and applying some of the principles and constructs used in this study could be the basis for similar studies of project managers in other industries.

This research could be extremely important to those who seek to examine and understand the role of formal learning, informal learning and its subset incidental learning in the workplace. However, based on the research, I would suggest that initially, formal learning (instructor-led or planned events), coupled with informal learning (mentoring, self-directed learning, and networking), followed by incidental learning (the actions of others, learning from involvement, and learning from mistakes), appears to be the best learning approach for those that want to initiate learning in a structured learning environment.

Additionally, I believe this study brought out and highlighted the specific activities that characterize each of the aforementioned modalities of learning so that project managers in the EPCM industry can be more informed about the methods that are available to them and, as a result, be more effective in selecting and pursuing a CPE agenda that will give them the return on investment they desire.

Implications for Practice

Companies that have adopted project management and use it as a tool to help achieve their goals and objectives understand the importance of employing experienced, competent project managers who can deliver projects on time, under budget, within quality standards, and with minimal deviation from scope.

Many times, regardless of where these professionals are geographically, recruiters working on behalf of companies will contact them and inquire about their availability and willingness to relocate. For example, I was contacted in February - 2017, by a recruiter who offered me a project manager position in Billings, Montana. Because I live in the Chicagoland area, that opportunity is not something I would be interested in. In addition, similar opportunities have been discussed in the first quarter of 2018.

The implication of this study for the EPCM industry could be far reaching, in that it suggests the need for a structured training program that combines formal learning, informal learning, and mentoring initially, while advocating PMP certification and CPE later on. Strategically, presidents and CEOs of companies in the EPCM industry should reconsider their positions relative to project management training and make project management training a priority. No longer should training be considered an afterthought and secondary to billable hours, especially when data from the Standish Group indicates a shortfall in the success rate of project delivery.

I recommend that EPCM companies, in addition to formal learning, informal learning and mentoring, consider project managers experts if they have completed a reasonable number of projects successfully. In addition, those that have 10 or more years of experience gained through the "school of hard knocks" or "learned commonsense" should be considered candidates for the title of expert project manager. Moreover, all EPCM companies should adopt an aggressive PMP program which requires all project managers to pursue and achieve PMP certification. Once this is achieved, it will require project managers to embark on a CPE agenda that will lead to lifelong learning, the results of which, I believe, will lead to a higher success rate of project delivery.

Suggestions for Future Research

There were several areas for potential research that emerged from this study. They exist, in part, because of my conscious decision not to do research in these areas. The first two areas listed below are indigenous to the EPCM industry and could be of interest to companies that seek to learn about the experiences of minority project managers in the U.S. All five of the areas listed below could be of interest to providers of project management services, customers of project management services, colleges and universities, and project management educators who seek a basis for future research.

1. How does achieving PMP certification early in one's career impact their progression from novice to expert?
2. How does making a conscious effort to train project managers impact a company's success in delivering projects in the EPCM industry?
3. How does not having a mentor impact the development of a project manager?
4. How long (in terms of years) does it take a project manager to achieve expert status with the benefit of formal and informal training?
5. How does a project manager in information technology maintain his or her level of expertise?

These questions can be asked in other industries such as healthcare, real estate, banking, marketing, sales, and human resources. In addition, a study could be crafted in such a way that different age groups are examined along with marital status and educational background. The point is that project management permeates and impacts all industries.

If we hope to improve, we must begin to ask ourselves very specific questions: 1) Are we doing project management effectively? 2) Are we doing it efficiently? If the answer is no, then the question that follows should include the words, 3) How do we improve.

REFERENCES

Alexander, I.F., & Stevens, R. (2002). *Writing better requirements.* London, UK: Addison-Wesley.

Anderson, J. R. (Ed.). (1981). *Cognitive skills and their acquisition.* Hillsdale, NJ: Erlbaum.

Association for Project Management. (2000). *APM body of knowledge* (5th ed.). Buckinghamshire, UK: Author.

Azzaretto, J. F. (1990) Power, responsibility, and accountability in continuing professional education. In R. M. Cervero & J. F. Azzeretto (Eds.), *Visions for the future of continuing professional education* (pp 25-50). Athens, GA: Department of Adult Education, University of Georgia.

Benner, P. (1982). From novice to expert. *American Journal of Nursing, 82*(3), 402-407.

Benner, P. (1983). Uncovering the knowledge embedded in practice. *Image: The Journal of Nursing Scholarship, 15*(2), 36-41.

Benner, P. (1984). *From novice to expert: Excellence and power in clinical nursing practice.* Menlo Park, CA: Addison-Wesley.

Benner, P., & Tanner, C. (1987). Clinical judgment: How expert nurses use intuition. *American Journal of Nursing, 87*(1), 23-31.

Bierema, L. L., & Eraut, M. (2004). Workplace-focused learning: Perspectives from HRD and CPE. *Advances in Developing Human Resources, 6*(1), 52-68.

Bloom, B.S. (Ed.). (1985). *Developing talent in young people.* New York, NY: Ballantine.

Brown, J. S., & Thomas, D. (2006). You play World of Warcraft? You're hired. *Wired Magazine, 14*(1), 1-3. Retrieved from http://www.wired.com/wired/archive/ 14.04/learn.html

Brown, K. L. (2000). Analyzing the role of the project consultant: Cultural change implementation. *Project Management Journal, 31*(3), 52.

Chase, W. G. (Ed.). (1973). *Visual information processing.* New York, NY: Academic Press.

Cheetham, G., & Chivers, G. E. (2005). *Professions, competence, and informal learning.* Cheltenham, UK: Edward Elgar Publishing.

Chi, M. (2006). Two approaches to the study of experts' characteristics. In K. A. Ericsson, N. Charness, P. J. Feltovich, & R. R. Hoffman (Eds.), *Cambridge handbook of expertise and expert performance* (1st ed., pp. 21-30). Cambridge, England: Cambridge University Press.

Chi, M.T.H., Feltovich, P.J., & Glaser, R. (1980). Categorization and representation of physics problems by experts and novices. *Cognitive Science, 5*(2), 121-152.

Chi, M. T. H., Glaser, R., & Farr, M. J. (Eds.). (1988). *The nature of expertise.* Hillsdale, NJ: Erlbaum.

Clarke, N. (2010). Projects are emotional: How project managers' emotional awareness can influence decisions and behaviours in projects. *International Journal of Managing Projects in Business, 3*(4), 604-624.

Colvin, G. (2008). *Talent is overrated.* New York, NY: Penguin.

Contreras, A. (2007). I'm an expert and you can be one too. *The Chronicle Review, 53*(45), B5. Retrieved from http://www.Chronicle.com

Creswell, J.W. (2005). *Educational research: Planning, conducting, and evaluating quantitative and qualitative research.* Upper Saddle River, NJ: Pearson Prentice Hall.

Daley, B. J. (1999). Novice to expert: An exploration of how professionals learn. *Adult Education Quarterly, 49*(4), 133-147.

Darkenwald, G. G., & Merriam, S. B. (1982). *Adult education: Foundations of practice.* New York, NY: Harper & Row.

Davison, T. (1997). *Equivalence and the recognition of prior learning in universities.* Unpublished manuscript, James Cook University, Queensland, Australia.

DeCarlo, D. (2004). *Extreme project management.* San Francisco, CA: Jossey-Bass.

Dewey, J. (1938). *Experience and education.* New York, NY: Collier Books.

Dreyfus, H., & Dreyfus, S. (1985). *Mind over machine: The power of human intuition and expertise in the era of the computer.* New York, NY: Free Press.

Dreyfus, S. E., & Dreyfus, H. L. (1980). *A five-stage model of the mental activities involves in directed skill acquisition.* Unpublished report supported by the Air Force Office of Scientific Research, No. Contract F49620-79-C-0063. University of California at Berkley.

DuBois, M., Koch, J., Hanlon, J., Nyatuga, B., & Kerr, N. (2015). Leadership styles of effective project managers: Techniques and traits to lead high performance teams. *Journal of Economic Development, Management, IT, Finance, & Marketing, 7*(1), 30-46.

Ekstedt, E., Lundin, R. A., Soderholm, A., & Wirdenius, H. (1999). *Neo-industrial organizing: Renewal by action and knowledge formation in a project-intensive economy.* London: Routledge.

Eraut, M. (1994). *Developing professional knowledge and competence.* London: Falmer Press.

Eraut, M. (2004). Informal learning in the workplace. *Studies in Continuing Education, 26*(2), 247-273.

Ericsson, K. A. (1996). *The road to excellence: The acquisition of expert performance in the arts and sciences, sports and games.* Mahwah, NJ: Erlbaum.

Ericsson, K. A. (2005). Recent advances in expertise research: A commentary on the contributions to the special issue. *Applied Cognitive Psychology, 19*(2), 233-241.

Ericsson, K. A. (2006). The influence of experience and deliberate practice on the development of superior expert performance. In K. A. Ericsson, N. Charness, P. Feltovich, & R. Hoffman (Eds.), *The Cambridge handbook on expertise and expert performance* (1st ed., pp. 683-703). Cambridge, England: Cambridge University Press.

Ericsson, K. A., & Smith, J. (Eds.). (1991). *Toward a general theory of expertise: Prospects and limits.* Cambridge, England: Cambridge University Press.

Expert. (1968). *Webster's New World Dictionary.* Cleveland, OH: World Publishing.

Expertise. (1999). *Merriam Webster's Dictionary.* Springfield, MA: Author.

Feltovich, P. J., Ford, K. M., & Hoffman, R. R. (Eds.) (1997). *Expertise in context: Human and machine.* Cambridge, MA: AAAi/MIT Press.

Galton, F. (1869/1979). *Hereditary genius: An inquiry into its laws and consequences.* London, UK: Julian Friedman.

Garrick, J. (1998). *Informal learning in the workplace.* London: Routledge.

Gilley, J. W., & Cunich, A. M. (1998). *Partnering to maximize organizational performance.* Cambridge, UK: Perseus.

Gray, C. F., & Larson, E. W. (2006). *Project management: The managerial process.* New York, NY: McGraw Hill/Irwin.

Grochow, D. (2008). From novice to expert: Transitioning graduate nurses. *Nursing Management, 39*(3), 10-12.

Hager, P. (1998). Recognition of informal learning: Challenges and issues. *Journal of vocational education and training, 50*(4), 521-535.

Hoffman, R. R. (Ed.). (1992). *The psychology of expertise: Cognitive research and empirical AI.* New York, NY: Springer-Verlag.

Holyoak, K. (1991). Symbolic connectionism: Toward third generation of theories of expertise. In K. A. Ericsson & J. Smith (Eds.), *Toward a general theory of expertise: Prospects and limits* (pp. 301-355). Cambridge, England: Cambridge University Press.

Houle, C. O. (1980). *Continuing learning in the professions.* San Francisco, CA: Jossey-Bass.

Hrimech, M. (2005). Informal learning. In L. M. English (Ed.), *International encyclopedia of adult education* (pp. 310-312). New York: Palgrave Macmillan.

Illich, I. (1977). Disabling professions. In I. Illich, I. K. Zola, J, McKnight, J. Caplan, & H. Shaiken (Eds.), *Disabling professions* (pp. 11-39). London, UK: Marion Boyars.

Imel, S., Brockett, R. G., & James, W. B. (2000). Defining the profession: A critical appraisal. In A. L. Wilson & E. R. Hayes (Eds.), *Handbook of adult education and continuing education* (pp. 628-642). San Francisco, CA: Jossey-Bass.

Jeris, L. H. (2010). Continuing professional education. In C.E. Kasworm, A.D. Rose, & J. M. Rose-Gordon (Eds.), *Handbook of adult and continuing education* (pp. 275-284). Thousand Oaks, CA: Sage.

Jeris, L. H. & Armoacost, L. K. (2002). Doing good or doing well? A counter-story of continuing professional education. *Learning in Health and Social Care, 1*(2), 94-104.

Kerzner, H. (2009). *Project management: A systems approach to planning, scheduling, and controlling* (10th Ed.) (p. 2). Hoboken, NJ: John Wiley & Sons.

King, P. K. (2010). *Informal learning in a virtual environment.* In C.E. Kasworm, A.D. Rose, & J. M. Rose-Gordon (Eds.), *Handbook of adult and continuing education* (pp. 421-430). Thousand Oaks, CA: Sage.

Knowles, M. (1970). *The modern practice of adult education: Andragogy vs. pedagogy.* New York, NY: Associated Press.

Kumar, S., & Shah, L. (2006). Evaluating the effectiveness of CPD and measuring returns on investment: A case study of UGC refresher courses. In P. Genoni & G. Walton (Eds.), *Continuing professional development – Preparing for new roles in libraries: A voyage of discovery.* Berlin, Germany: K. G. Saur.

Labelle, T. J. (1982). Formal, non-formal, and informal education: A holistic perspective on lifelong learning. *International Review of Education, 28*(2), 158-175.

Larson, M. E. (1984). Expertise and expert power. In T.L. Haskell (Ed.), *The authority of experts.* Bloomington: Indiana University Press.

Lindeman, E.C. (1926). *The meaning of adult education,* New York, NY: New Republic.

Lisman, D. (1980). Mandatory continuing education and teachers. *Phi Delta Kappan, 62*(2), 125-126.

Marsick, V. J., & Watkins, K. (1990). *Informal and incidental learning in the workplace*. London, UK: Routledge.

Merriam, S. B. (1998). *Qualitative research and case study application in education* (2nd ed.) San Francisco, CA: Jossey-Bass.

Merriam, S. B. (2002). *Qualitative research in practice*. San Francisco, CA: Jossey-Bass.

Merriam, S.B., Caffarella, R.S., & Baumgartner, L. M. (2007). *Learning in adulthood: A comprehensive guide* (3rd ed.). San Francisco, CA: Jossey-Bass.

Mocker, D.W., & Spear, G.E. (1982) Lifelong learning: Formal, non-formal, informal, and self-directed. Retrieved from ERIC database. (ED220723)

Mumford, A. (1995). Four approaches to learning from experience, *Industrial and Commercial Training, 27*(8), 12-19.

Munk-Madsen, A. (2005, October). *Define project*. Paper presented at the 28th Information Systems Research Seminar in Scandinavia, Kristiansand, Norway. Retrieved from http://wwwold.hia.no/iris28/Docs/IRIS20281039.pdf

National Center for Education Statistics. (2007) *Adult education participation in 2004-2005*. Retrieved from http://nces.ed.gov/pubs2006/adulted/

Newell, A., & Simon, H.A. (1972). *Human problem solving*. Englewood Cliffs, NJ: Prentice Hall.

Nicholas, J. M., & Steyn, H. (2008). *Project management for business, engineering, and technology* (3rd Ed.). Burlington, MA: Elsevier.

Nowlen, P. M. (1988). *A new approach to continuing education for business and the professions*. New York, NY: MacMillan.

Oberg, R., Probasco, L., & Ericsson, M. (2003). *Applying requirements management with use cases.* Rational Software Corporation. Retrieved from www.compgraf.ufu.br/ alexandre/esof/ use_cases.pdf

Ohliger, J. (1981). Dialogue on mandatory continuing education. *Lifelong learning: The adult years, 4*(10), 5, 7, 24-26. Retrieved from ERIC database. (EJ246684)

Padgett, C. (2009). *The project success method: A proven approach for achieving superior projects.* Hoboken, NJ: John Wiley.

Pandya, K. D. (2014). The key competencies of project leader beyond the essential technical capabilities. *IUP Journal of Knowledge Management, 12*(4), 39-48.

Project Management Institute. (2004). *A guide to the project management body of knowledge.* (4th ed.). Newton Square, PA: Author.

Project Management Institute (2005). *A guide to the project management body of knowledge.* (5th ed.). Newton Square, PA: Author.

Queeney, D. S. (1996) Continuing professional education. In R. L. Craig (Ed.), *The ASTD training and development handbook* (4th ed., pp. 698-724). New York, NY: McGraw-Hill.

Queeny, D.S. (2000). Continuing professional education. In A.L. Wilson & E. R. Hayes (Eds.), *Handbook of adult education and continuing education* (pp. 375-391). San Francisco, CA: Jossey-Bass.

Robbins, S. P., & Coulter, M. (2012). *Management* (11th ed.). Upper Saddle River, NJ: Pearson.

Rockhill, K. (1983). Mandatory continuing education for professionals: Trends and issues. *Adult Education Quarterly, 33*(2), 106-116.

Schon, D.A. (1983) *The reflective practitioner*: New York, NY: Basic Books.

Shen, G.Q., & Yu, A. T. (2013). Problems and solutions of requirements management for construction projects under the traditional procurement systems. *Facilities, 31*(5/6), 223-237.

Simon, H. A., & Chase W. G. (1973). Skill in chess. *American Scientist, 61*(4), 394-403.

Simonton, D. K. (1977). Creative productivity, age, and stress: A biographical time series analysis of 10 classical composers. *Journal of Personality and Social Psychology, 35(11),* 791-804.

Stake, R. E. (1995). *The art of case study research.* Thousand Oaks, CA: Sage.

Standish Group International. (2004). *2004 third quarter research report.* Retrieved from http://www.standishgroup.com

Standish Group International. (2009). *2009 annual research report.* Retrieved from http://www.standishgroup.com

Starkes, J., & Allard, F. (Eds.). (1993). *Cognitive issues in motor expertise.* Amsterdam, The Netherlands: North Holland.

Starkes, J., & Ericsson, K. A. (Eds.). (2003). *Expertise performance in sport: Recent advances in research on sport exercise.* Champaign, IL: Human Kinetics.

Tanner, C., Padrick, K., Westfall, U., & Putzier, D. (1987). Diagnostic reasoning strategies for nurses and nursing students. *Nursing Research, 36*(6), 358-363.

Turner, J. (1993). *Handbook of project-based management: Improving the process for achieving strategic objectives.* London: McGraw-Hill.

Walker, G. (1997). Find the right consultant in seven steps. *Quality World, 23*(6), 470-472.

White, D. S. (2015). Novice to expert: Mentoring the graduate nurse. *Tennessee Nurse, 78*(1), 12.

Wilkinson, S. (2001). An analysis of the problems faced by project management companies managing construction projects.

Engineering, Construction, and Architectural Management, 8(3), 160-170.

Woolls, B. (2006). Continuing professional education to continuing professional development and workplace learning: The journey and beyond. In P. Genoni & G. Walton (Eds.), *Continuing professional development – Preparing for new roles in libraries: A voyage of discovery*. Berlin, Germany: K. G. Saur.

Wysocki, K., Lewis, J., & DeCarlo, D. (2001). *The world-class manager: A professional development guide*. Cambridge, UK: Perseus.

Appendix A

INTERVIEW QUESTIONS

Interview Questions

Novice PM Questions

1. What prompted you to become a project manager (PM)?

    a. How would you describe the process of your growth from your beginning as a project manager to now?
    b. How did you learn what a PM does and how to do it?
    c. How did you learn to solve the problems that you have faced so far as a PM?
    d. What is the most important thing you have learned through your work as a PM?

2. What events or activities had the greatest impact on your professional development so far as a PM?

    a. For each event, describe what happened
    b. For each event describe the impact on your development

3. Who are the people who have the greatest impact on your professional development as a PM? (Teachers, students, colleagues …)

    a. Describe your interaction with each person
    b. Describe the impact of each person on your professional development

4. How would you describe the atmosphere in your current department? What influence has the department had on your professional development?

5. Is there anything outside of your professional life that has particularly affected your development as a PM?

6. Reviewing your start in the PM profession, what advice do you have for people getting into the field of PM
7. Reviewing your life as a PM, what advice do you have for peers or coworkers that are at your level in the field of project management?
8. Is there anything else you want to tell me about your development as a PM? Is there anything specific that you do to sharpen or maintain your craft?

Intermediate and Expert PM Questions

1. What prompted you to become a project manager (PM)?
   a. How would you describe the process of your growth from novice to present?
   b. How did you learn what a PM does and how to do it?
   c. How did you learn to solve the problems a PM faces?
   d. What is the most important thing you have learned through your work as a PM?
2. What events or activities had the greatest impact on your professional development as a PM?
   a. For each event, describe what happened
   b. For each event describe the impact on your development
3. Who are the people who have the greatest impact on your professional development as a PM? (Teachers, students, colleagues …)
   a. Describe your interaction with each person

    b. Describe the impact of each person on your professional development
4. How would you describe the atmosphere in your current department? What influence has the department had on your professional development?
5. Is there anything outside of your professional life that has particularly affected your development as a PM?
6. Reviewing your start in the PM profession, what advice do you have for people getting into the field of PM
7. Reviewing your life as a PM, what advice do you have for novices
8. Is there anything else you want to tell me about your development as a PM? Is there anything specific that you do to sharpen or maintain your craft?

Appendix B

INVITATION TO PARTICIPATE

Invitation to Participate

Date

Dear_____:

I am a "Researcher" and I would like to invite you to participate in my research study titled, "Project Management: Novice to Expert: An Examination of the Engineering Procurement, and Construction Management (EPCM) Industry." As a Researcher, I am interested in learning more about the road map to project management expertise. Specifically, what steps should a new project manager take to become an expert in the field of project management?

I aim to gain insight on this topic through face to face one-on-one interviews. This letter is being sent to project managers who currently work in the EPCM industry and have managed projects for at-least 1year and not more than 30 years. All interviews will take place at the participants' companies, or a place designated by the participant; the interviews will last approximately 30-45 minutes. Interviews will be digitally recorded with a voice recorder, and later transcribed. I will ask you to review the transcript for accuracy. To ensure confidentiality, I will refer to you by using a pseudonym rather than your real name. Furthermore, participation in this study will in no way impact your career status at your place of employment. Participation is strictly voluntary and you may withdraw from the study at any time without any penalty whatsoever.

If you do decide to participate in this research study, are chosen to participate, and complete the interview, you will receive a $$$ gift card to Starbucks. If you are interested in participating, please complete the demographic sheet attached to this email and send it directly to me at yahoo.com.

Thank you for your consideration and please do not hesitate to contact me if you have questions.

Sincerely,

Derrick J. Walters
Email:
Phone:

Appendix C

INFORMED CONSENT FORM

How do Project Managers Become Experts:

An Examination of Project Managers in the EPCM Industry

Derrick J. Walters

PARTICIPANT CONSENT FORM

Dear Participant:

I am a Researcher and I would like to invite you to participate in my research study titled, "Project Management: Novice to Experts: An Examination of the Engineering Procurement, and Construction Management (EPCM) Industry." Please sign your name under the "Authorization" section, section A, if you agree to participate.

A. Authorization

I,_____ (Participant's Name), hereby consent to participate in a study which will involve an interview (in person) performed by Derrick J. Walters.

B. Description

o The title of the study is: "How do Project Managers Become Experts: An Examination of The EPCM Industry"
o This study involves a personal interview about my perceptions and experiences working as a project manager in the EPCM industry, and my journey from novice project manager to expert project manager.
o The Interviews are tentatively scheduled for "Date", and they will be held at a time and place of my choosing; they will take approximately 45-60 minutes."

- The purpose of this study is to gain a better understanding of how project managers become experts, examining their journey from novice project manager to expert project manager. Findings of this study will provide new insights regarding project managers and their journey to expert, as well how they maintain the expert status and work at peak performance.
- As a participant, I will be interviewed and asked several questions about project management in the context of the EPCM industry. The interview will be digitally recorded (voice recording). I understand that I may elect not to answer any questions if I so desire. I also understand that I may terminate the interview at any time without any penalty.
- All data will be kept confidential and my name will not be divulged in a verbal or written manner.
- I will provide the researcher with a pseudonym to ensure my confidentiality. I understand that the researcher can use my demographic information; however, if I request, the researcher will not include any specified information in a research report.
- Once the digital recording (voice recording) is transcribed, the researcher will modify and code the names included in the transcripts. Once the study is completed the recording and coding sheet will be destroyed.
- If I do decide to participate in this research study, am chosen to participate, and complete the interview, I will receive a $$$ gift card to Starbucks.
- If I have any questions about the research study, If I have any questions about my

Derrick can be contacted:
Derrick Walters

C. **Benefits**

The researcher believes that understanding how project managers become experts will enhance the number of project managers who transition from novice to expert, enable them to maintain their expert status and lay the foundation for similar studies of project managers in other industries.

D. **Voluntary Participation**

I understand that participation is voluntary and that I will not be penalized if I choose not to participate. I also understand that I am free to withdraw my consent and end my participation in this study at any time without penalty, after I notify the researcher.

E. **Consent**

I have read and fully understand the consent form. I sign it freely and voluntarily. I have received a copy of this form.

Date:_____

Time:_____(a.m./p.m.)

Signature of participant:_____

I agree to be digitally recorded (voice recorded) during the interview:

Signature of participant:_____

Please provide the researcher your signed consent form and remember to keep the other copy of this form for your records.

Appendix D

DEMOGRAPHICS FORM

Demographics Form

Age: _____ Gender _____ College Major: _____

Ethnicity: White ☐ Black ☐ Hispanic ☐ Asian or Pacific Islander ☐ American Indian o Other o please explain _____

Do you have formal project management education? If so, at which college?

_____
_____
_____

Do you have project management experience? If so, how many years' experience?

_____
_____
_____

How many projects have you completed successfully, On-time, under-budget, and within quality standards?

_____
_____
_____

Have you participated in continuing professional education (CPE), or any informal learning that you are aware of?

_____
_____
_____

Do you consider yourself a beginning project manager (novice), fairly experienced (mid-career) or an accomplished project manager (project management expert)?

_____
_____
_____

Are you a member of the Project Management Institute (PMI)? If so, how many years have you been a member?

_____
_____
_____

Appendix E

CODE-TO-THEME TABLES

Study 1 - Brad

Key Word(s)	Sample Quote	Phrase(s)	Theme
Project Management	Brad said he migrated toward project management	Brad did not have a project management mentor	Project Management
Prof Engineering License	Pursuit and maintenance of a PE license has a dramatic impact on his life	Brad has to maintain a CPE agenda by acquiring 30 PDUs every 3 years	Continuing Professional Education
Self-direction	I was a jack or all trades	I was responsible for everything	Informal Learning
Learned from Involvement	I watched another project manager	Brad learned from mistakes	Incidental Learning
Project management Professional (PMP)	I took a 1-day course in Project Management – 20 years ago	After I pass the PMP exam I will take more project management courses	Formal Learning
Advice	Find a mentor	Take a class, read books	Open Theme

Study 2 – Karen

Key Word(s)	Sample Quote	Phrase(s)	Theme
Project management	Recruited into project management	Her current company talked to her about project management when she graduated from college	Project Management
CPE	CPE kind of happened throughout my career	CPE played an important role in my learning process	Cont. Prof Ed
Mentor	I was mentored as a project engineer	I was self-directed as a project manager	Informal Learning
Colleagues	I develop relationships with others - Networking	I believe in learning by doing, and asking questions	Incidental Learning
Project management Professional (PMP)	I've been considering PMP certification	Class room training was important	Formal Learning
Advice	Get some formal training, pursue PMP certification	Build relationships and get as much experience as possible	Open Theme

Study 3 – Jeffrey

Key Word(s)	Sample Quote	Phrase(s)	Theme
Project management	I was hired by a project management firm when I graduated from college	I received training on project management processes, the role of a project team, the role of a project manager etc.	Project Management
Learning MS Project	I thought MS Project was a burden until I learned how to use it	I learned to predict if my project would finish at a certain time using MS Project	Continuing Professional Education
Self-paced study guides	I learned through Self-directed study how facilitation skills could benefit me	I applied the self-directed learning techniques and saw an improvement in my meetings	Informal Learning
Involvement and mistakes	I failed to define when an activity was considered complete	I complicated the situation by taking another person's word, instead of investigating myself	Incidental Learning
Project management Professional (PMP)	I received my training from an REP (Registered Education Provider)	I received project champion training, as well as project manager training	Formal Learning

| Advice | Be willing to step outside of your comfort zone | Get Project management training and experience through: Internships, The Boys and Girls club, Charity work, and the United way | Open Theme |

Appendix F

MEMBER CHECK - FOLLOWUP EMAIL

I used convenient sampling to select project managers for this study. In the process I contacted four project managers, but only three agreed to participate.

Brad - Member Check Email

Shown below is the email I sent Brad to get his feedback on the interviewing process. Brad provided his follow-up via telephone and said the transcripts accurately reflected his comments in the interview.

From: "D.Walters, PMP" <>\
To:
Sent:
Subject: "DWalters_Dissertation_Followup"

Hello Brad:

Attached is a transcript of our interview; it came back yesterday. Please review it for accuracy and try to get back to me with comments by Friday of this week.

Thanks

PS ... Please call me if you have questions
c:

Derrick

Derrick J. Walters, MBA, PMP®, ITIL, EdD
Professor and Project Management Consultant
Adjunct Professor NLU, DeVry, UOP
Pres. & Prin. Consultant
Walters Consulting, LLC
Bus Ph:
Fax:
Website: www.wc-llc.com

I used convenient sampling to select project managers for this study. In the process I contacted four project managers, but only three agreed to participate.

Karen - Member Check Email

Shown below is the email I sent Karen to get her feedback on the interviews conducted, by me, for this study. Karen provided her follow-up via telephone and said the transcripts accurately reflected her comments in the interview.

From: "D.Walters, PMP" <>
To:
Sent:
Subject: "DWalters_Dissertation_Followup"

Hello Karen:

Attached is a transcript of our interview; it came back yesterday. Please review it for accuracy and try to get back to me with comments by Friday of this week.

Thanks

PS ... Please call me if you have questions
c:

Derrick

Derrick J. Walters, MBA, PMP®, ITIL, EdD
Professor and Project Management Consultant
Adjunct Professor NLU, DeVry, UOP
Pres. & Prin. Consultant
Walters Consulting, LLC
Bus Ph:
Fax:
www.wc-llc.com
dwalters@wc-llc.com

I used convenient sampling to select project managers for this study. In the process I contacted four project managers, but only three agreed to participate.

Jeffrey - Member Check Email

Shown below is the email I sent Jeffrey to get his feedback on the interviews conducted, by me, for this study. Jeffrey provided his follow-up via telephone and said the transcripts accurately reflected his comments in the interview.

From: "D.Walters, PMP" <>
To:
Sent:
Subject: "DWalters_Dissertation_Followup"

Hello Jeffrey:

Attached is a transcript of our interview; it came back yesterday. Please review it for accuracy and try to get back to me with comments by Friday of this week.

Thanks

PS … Please call me if you have questions
C :

Derrick

Derrick J. Walters, MBA, PMP®, ITIL, EdD
Professor and Project Management Consultant
Adjunct Professor NLU, DeVry, UOP
Pres. & Prin. Consultant
Walters Consulting, LLC
Bus Ph:
Fax:
Website: www.wc-llc.com